Intelligent Surveillance Systems

International Series on
INTELLIGENT SYSTEMS, CONTROL AND AUTOMATION:
SCIENCE AND ENGINEERING

VOLUME 51

For other titles published in this series, go to
www.springer.com/series/6259

Huihuan Qian · Xinyu Wu · Yangsheng Xu

Intelligent Surveillance Systems

Springer

Huihuan Qian
Mechanical and Automation Engineering
The Chinese University of Hong Kong
Shatin, New Territory, Hong Kong SAR
China, People's Republic
hhqian@mae.cuhk.edu.hk

Yangsheng Xu
Mechanical and Automation Engineering
The Chinese University of Hong Kong
Shatin, New Territory, Hong Kong SAR
China, People's Republic
ysxu@cuhk.edu.hk

Xinyu Wu
Shenzhen Institute of Advanced Technology
Xueyuan Road 1068
518055 Shenzhen
China, People's Republic
xy.wu@siat.ac.cn

ISBN 978-94-007-3560-6 ISBN 978-94-007-1137-2 (eBook)
DOI 10.1007/978-94-007-1137-2
Springer Dordrecht Heidelberg London New York

Cover design: SPi Publisher Services

Printed on acid-free paper

Springer is part of Springer Science+Business Media (www.springer.com)

To our families

Preface

Surveillance systems have become increasingly popular in the globalization process. However, the full involvement of human operators in traditional surveillance systems has led to shortcomings, for instances, high labor cost, limited capability for multiple-screen monitoring, inconsistency during long-durations, etc. Intelligent surveillance systems (ISS) can supplement or even replace traditional ones. In ISSs, computer vision, pattern recognition, and artificial intelligence technologies are developed to identify abnormal behaviors in videos. As a result, fewer human observers can monitor more scenarios with high accuracy.

This book presents the research and development of real-time behavior-based intelligent surveillance systems, at the Chinese University of Hong Kong, as well as Shenzhen Institute of Advanced Technology (Chinese Academy of Sciences). We mainly focus on two aspects: 1) the detection of individual abnormal behavior based on learning; 2) the analysis of dangerous crowd behaviors based on learning and statistical methodologies. This book addresses the video surveillance problem systematically, from the foreground direction, blob segmentation, individual behavior analysis, group behavior analysis, unsegmentable crowd behavior analysis.

This book is appropriate for postgraduate students, scientists and engineers with with interests in the computer vision, and machine intelligence. This book can also serve as a reference book for algorithms and implementation in surveillance.

We would like to thank Dr. Yongsheng Ou, Dr. Weizhong Ye, Dr. Zhi Zhong, Dr. Xi Shi, Dr. Yufeng Chen, Dr. Guoyuan Liang, Dr. Shiqi Yu, Mr. Ning Ding, and Mr. Wing Kwong Chung for their support in the valuable discussion of this book. Thanks also go to Dr. Ka Keung Lee for his proofreading of the first draft.

Finally, this book is supported in part by Hong Kong Research Grant Council under CUHK4163/03E and National Natural Science Foundation of China under grant number 61005012.

The Chinese University of Hong Kong,
April 2011

Huihuan Qian
Xinyu Wu and
Yangsheng Xu

Contents

List of Figures

List of Tables

Chapter 1
Introduction

1.1 Background

With the development and globalization of human social activities, surveillance systems has become increasingly important and popular in public places such as banks, airports, public squares, casinos. A significant amount of cameras (Figure 1.1) are been installed daily to monitor the public areas or private areas for the sake of security and safety.

In traditional surveillance systems, human operators are employed to monitor activities that are captured on video. They extend the vision range of securities, and reduce the security manpower cost, and enhance the security efficiency to some extend. However, such systems have obvious disadvantages, including the high cost of human labor, the limited capability of operators to monitor multiple screens, inconsistency in long-duration performance, and so forth. Intelligent surveillance systems (ISSs) can supplement or even replace traditional systems. In ISSs, computer vision, pattern recognition, and artificial intelligence technologies are used to identify abnormal behaviors in videos and thus potentially in real time.

This book presents the development of a real-time human behavior-based surveillance system at the Chinese University of Hong Kong. We focus on two fields in the intelligent video surveillance research: the detection of individual abnormal behavior and the analysis of crowd behavior. For abnormal behavior detection, we apply human behavior learning and modeling methods to identify the abnormal behavior of individuals in different environments. For crowd modeling, we adopt the texture analysis and optical flow methods to analyze human behavior in crowds and potentially dangerous crowd movements. Several practical systems are introduced, which include a real-time face classification and counting system, a surveillance robot system that utilizes video and audio information for intelligent interaction, and a robust person counting system for crowded environments.

H. Qian et al., *Intelligent Surveillance Systems*, Intelligent Systems, Control and Automation: Science and Engineering 51, DOI 10.1007/978-94-007-1137-2_1,

Fig. 1.1 Surveillance cameras

1.2 Existing Surveillance Systems

Many different types of video surveillance systems are available worldwide, covering different aspects of surveillance. We do not provide a comprehensive survey of those systems but do highlight a number of significant systems.

W^4 is a real-time visual surveillance system for detecting and tracking multiple persons and monitoring their activities in an outdoor environment. It operates on monocular gray-scale video imagery, or on video imagery from an infrared camera. The system employs a combination of shape analysis and tracking to locate people and parts of their body (head, hands, feet, torso) to create models of people's appearance so that they can be tracked through interactions including those involving occlusion. It runs at 25 Hz for 320 x 240 pixel images on a 400 MHz Dual Pentium II PC.

PFinder is a real-time system for tracking people and interpreting their behavior. It runs at 10 Hz on a standard SGI Indy computer, and has performed reliably on thousands of people in many different physical locations. The system uses a multiclass statistical model of color and shape to obtain a two-dimensional (2D) representation of head and hands in a wide range of viewing conditions. Pfinder has successfully been used in a broad range of applications including wireless interfaces, video databases, and low-bandwidth coding.

Zhao [164] proposed a system that can track multiple individuals in dynamic scenes. Traditional methods rely on appearance models that must be acquired when humans enter a scene and are not occluded. Zhao's system can track humans in crowded environments with significant and persistent occlusion by making use of both human shape and camera models. It is assumed that humans walk on a plane, and the camera acquires human appearance models.

1.3 Book Contents

This book aims to provide a comprehensive description of a novel multimodal surveillance system, including its algorithms, technology, and applications. The system can model individual, group, and crowd behavior. A sound recognition system is also introduced. Figure 1.2 illustrates the system architecture. The book is divided into eight chapters, including this introductory one.

In Chapter 2, we introduce the fundamental preprocessing of video frames, including background and foreground detection. The objective is to extract the foreground from the image. We consider two approaches: one is based on the fact that the background is stationary in the image captured by a fixed monocular camera, while the other is based on the assumption that the foreground contains objects of interest that are moving. The next sections elaborate the two approaches in detail. We outline the pattern classification, frame differencing, and optical flow methods in this chapter.

In Chapter 3, we cover 1) segmentation, to separate the merged foreground into the different image blobs of different persons, and 2) tracking, to give each blob a

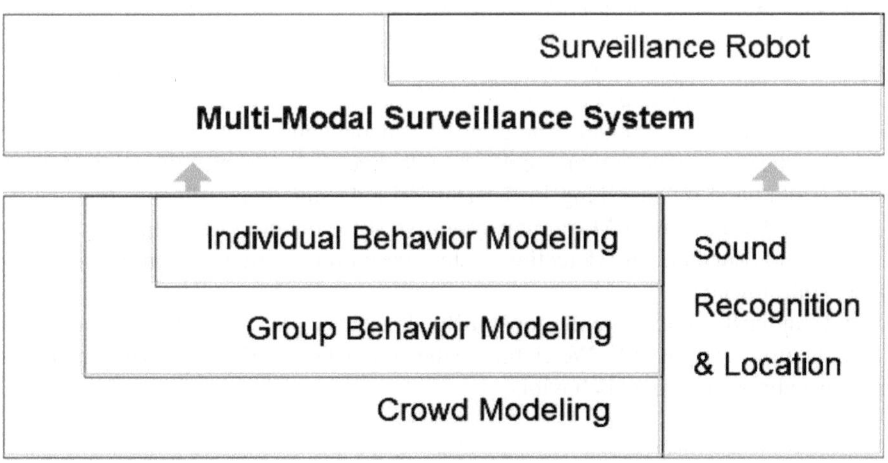

Fig. 1.2 System architecture

unique and correct identification (ID). This is necessary for the subsequent behavior analysis. Hybrid segmentation methods, including the histogram- and ellipse-based approaches, are elaborated, as well as tracking methods, including hybrid tracking (integrating distance and color tracking) and particle filter- and local binary pattern-based tracking.

After successful segmentation and tracking, the temporal-spatial information of each blob sequence can be indexed for further behavior analysis. In Chapter 4, we explain two kinds of approaches to the behavior analysis of individuals: learning-based and rule-based. In learning-based approaches, contour- and motion-based features are analyzed. A household surveillance robot is elaborated as an application.

In Chapter 5, a new real-time face classification system is presented. We classify facial images into two categories: Asian and non-Asian. Selected features of principal component analysis (PCA) and independent component analysis (ICA) are input into support vector machine (SVM) classifiers. The system can be used for other types of binary classification of facial images, such as gender and age classification, with only little modification.

In Chapter 6, we go one step further to consider multiple persons, i.e., human groups, in a field. A new multi-agent approach is proposed, which can be employed in the surveillance of groups in public places rather than tracking only individuals. The agent embodies the state and logic relationship between the person, whom the agent represents, and the other persons in the same group. The experimental results show that using our multi-agent approach to compute and analyze the relationships among a number of agents, we can efficiently perform real-time surveillance of a group and handle related events to enhance the applicability and intelligence levels of the surveillance system.

In Chapter 7, the monitored scenario is even more challenging; i.e., the environment is crowded, and tracking becomes difficult for the computer. We focus on human counting and distribution. A new learning-based method for people counting in a crowded environment using a single camera is introduced. The main difference between this method and traditional ones is that the former adopts separate blobs as the input into the people number estimator. Regarding human distribution, an approach to calculate a crowd density map using video frames is described. In the initial calibration stage, a set of multiresolution density cells is created based on a perspective projection model. The grey level dependence matrix (GLDM) feature vector is then extracted for each cell. A scale space is built, and the characteristic scale is obtained by searching in the Harris-Laplacian space. Finally, the feature vectors are fed into an SVM training system to solve the nonlinear regression problem. An abnormal density detection system based on multiresolution density cells is also introduced.

In Chapter 8, we explore the most difficult scenario, i.e., crowdedness such that segmentation is impossible. We elaborate the optical flow and energy-based methods to detect abnormal crowd behavior.

1.4 Conclusion

This book provides a comprehensive overview of the technologies and systems that we have developed in the area of intelligent surveillance. More and more surveillance cameras are being installed in different locations, including banks, government premises, railway stations, and houses. However, unsolved computer vision problems related to surveillance systems still exist, including behavior modeling in very complex environments, crowd behavior modeling, and multimodal surveillance analysis. Research directions that could potentially improve the performance of current systems include the following.

(1) Surveillance of crowded environments. Many theoretical and practical problems are encountered in crowd surveillance. Human motion in crowds is difficult to model, and innovative methods are required.

(2) Multimodal surveillance. From an application point of view, other media and information channels including audio ones could complement traditional video surveillance systems.

(3) Motion modeling under environmental constraints. Human motion is always related to the environment in which it occurs. Thus, incorporating the information related to the surveillance environment could improve the algorithms.

Given the great demand for surveillance applications, new intelligent video surveillance technology will dramatically improve the application of computer vision technology while providing new directions in theoretical modeling.

Chapter 2
Background/Foreground Detection[1]

2.1 Introduction

With the acquisition of an image, the first step is to distinguish objects of interest from the background. In surveillance applications, those objects of interest are usually humans. Their various shapes and different motions, including walking, jumping, bending down, and so forth, represent significant challenges in the extraction of foreground pixels from the image.

To tackle that problem, we consider two approaches: one is based on the fact that the background is stationary in the image captured by a fixed monocular camera, while the other is based on the assumption that the foreground contains objects of interest that are moving. The next sections elaborate on the two approaches in detail.

2.2 Pattern Classification Method

A popular detection methodology is based on the concept that once the background is extracted, the foreground can conveniently be obtained by subtracting the viewed image from the background model [162, 164].

2.2.1 Overview of Background Update Methods

Current background extraction methods include the multi-frame average, selection, selection-average, and random update methods [1–7] and Kalman filter-based method [8, 9]. These are briefly elaborated and compared below.

[1] Portions reprinted, with permission from Xinyu Wu, Yongsheng Ou, Huihuan Qian, and Yangsheng Xu, A Detection System for Human Abnormal Behavior, *IEEE International Conference on Intelligent Robot Systems*. ©[2005] IEEE.

H. Qian et al., *Intelligent Surveillance Systems*, Intelligent Systems, Control and Automation: Science and Engineering 51, DOI 10.1007/978-94-007-1137-2_2,

2.2.1.1 Multi-Frame Average Method

The multi-frame average method can be described as follows. For an image sequence $B_i \;\; i = 1,..,n$,

$$B_n = \frac{\sum_1^n B_i}{n} \tag{2.1}$$

Here, if the number of frames, n, is as large as needed, then B_n should be the background. However, in this method, too much memory is required to store the image sequence B_i. Hence, an approximate method is developed, which can be presented as

$$B_{pt} = kB_{pt-1} + (1-k)C_{pt-1} \qquad 0 < k < 1 \tag{2.2}$$

where

- B_{pt} —- a pixel in the current background;
- B_{pt-1} —- a pixel in the last background;
- C_{pt-1} —- a pixel in the current image; and
- k —- a threshold value.

The disadvantages of this method are: 1) the threshold value k is difficult to determine, and 2) k needs to be adjusted, depending on the degree of environmental change. When there are many objects in an active area, the noise in that area can be great, because the objects can also be deemed as background.

2.2.1.2 Selection Method

In this method, a sudden change in pixels is deemed to be the foreground, which will not be selected as the background.

$$\begin{aligned} &If \;\; |C_{p1} - C_{pt-1}| > T \\ &Then \;\; B_{pt} = B_{pt-1} \qquad (don't\ update) \\ &Else \;\; B_{pt} = C_{pt-1} \qquad (update), \end{aligned} \tag{2.3}$$

where

- B_{pt} —- a pixel in the current background;
- B_{pt-1} —- a pixel in the last background;
- C_{pt} —- a pixel in the current image of the image sequence;
- C_{pt-1} —- a pixel in the last image of the image sequence; and
- T —- a threshold value.

The disadvantage of this method is that determining the threshold value of T is difficult. If T is too small, then the background will be insensitive to change. If T is too large, then the background will be too sensitive to change.

2.2.1.3 Selection-Average Method

To address the disadvantages of the multi-frame average and selection methods, Fathy [19] proposed an improved method - the selection-average method - which is expressed as follows.

$$
\begin{aligned}
&If\ |C_{pt} - B_{pt}| < T_1 \\
&Then\ If\ |C_{pt} - C_{pt+1}| < T_2 \\
&Then\ \ B_{pt+1} = (B_{pt} + C_{pt+1})/2 \qquad (update) \\
&Else\ \ B_{pt+1} = B_{pt} \qquad (don't\ update),
\end{aligned}
\tag{2.4}
$$

where

- B_{pt+1} —- a pixel in the current background;
- B_{pt} —- a pixel in the last background;
- C_{pt+1} —- a pixel in the current image of the image sequence;
- C_{pt} —- a pixel in the last image of the image sequence; and
- T_1, T_2 —- threshold values.

2.2.1.4 Kalman Filter-based Adaptive Background Update Method

The Kalman filter-based model [8, 9] allows the background estimate to evolve as lighting conditions change with changes in the weather and time of day. In this method, the moving objects in the frame stream are treated as noise, and a Kalman filter is used to estimate the current background. However, it is obvious that in many cases, such as during rush hour or in a traffic jam, the moving object cannot simply be deemed noise. This method also requires much calculation time and is not discussed in detail here.

2.2.1.5 Another Adaptive Background Update Method

Another use of the adaptive background model can be found in [11]. The background equations are as follows:

$$
\begin{aligned}
A &= AND(M_{bkgnd}, I_{obj.mask}) \\
B &= AND\left(\frac{I_{current\ frame} + 15 \cdot M_{bkgnd}}{16}, NOT(I_{obj.mask})\right) \\
M_{bkgnd} &= OR(A, B),
\end{aligned}
\tag{2.5}
$$

where M_{bkgnd} is the estimated background, $I_{current\ frame}$ is the current image, and $I_{obj.mask}$ is a binary mask containing the foreground detected so far. $I_{obj.mask}$ is, in turn, computed as

$$I_{ojb.mask} = OR(I_{substraction_mask}, I_{optical_mask})$$

$$I_{substraction_mask} = \begin{cases} 0 & if\ |I_{current\ frame} - M_{bkgnd}| < K\sigma \\ 1 & otherwise \end{cases} \quad (2.6)$$

where $I_{substraction\_mask}$ is the binary mask of the foreground computed by background subtraction, and I_{optica_mask} is the foreground computed using optical flow estimation as discussed in the following paragraphs. The threshold used is a multiple (K) of the standard deviation σ of camera noise (modeled as white noise). K is determined empirically to be six.

2.2.1.6 Current Applications of Background Update Methods

In some complex cases, parts of the background are almost hidden behind an object; however, in the video stream, we can glimpse them. Figure 2.1 shows such an example. In the middle of this street is a place where people always stand, waiting for the signal. At that place, we can glimpse the background in only a few fragments of the video stream.

We capture an image sequence in the scene presented in the figure. To observe the result of the application of recent methods to this image sequence, we choose the multi-frame average and the selection-average method because the former is basic and the latter is popular. The results are shown in Figures 2.2-2.6. The selection-average method quickly eliminates moving objects. However, objects that move slowly cannot be removed completely. Another disadvantage of the selection-average method is that it needs an initial background. In this case, the initial background is difficult to capture. Here, the initial background in Figures 2.3-2.5 is set as Figure 2.2.

Figure 2.3 is almost the same as Figure 2.2. However, because $T1$ is small, the background cannot be updated correctly.

Figure 2.5 seems to strike a balance between updating and filtering with $T1 = 50$ and $T2 = 3$. Figure 2.6 uses a better initial background captured by the method proposed in the next section.

It can be seen that neither the average- nor the noise filter-based method works in complex scenarios, because the noise is too large to handle. Of course, any method that needs an initial background will not work, either.

In the next section, we propose a new method that can retrieve the background from a noisy video stream.

Fig. 2.1 People stay on street

Fig. 2.2 The Result Background of Multi-frame Average Method

2.2.2 Pattern Classification-based Adaptive Background Update Method

Definition: $\sim$

$$\begin{aligned} &For\ two\ n \times m\ images\ I_1,\ I_2. \\ &If\ \ |I_1(i,j) - I_2(i,j)| \leq T_1 \\ &Then\ \ I_1 \sim I_2, \end{aligned} \tag{2.7}$$

Fig. 2.3 The Result Background of Selection Average Method (T1 = 3 and T2 = 3)

Fig. 2.4 The Result Background of Selection Average Method (T1 = 15 and T2 = 9)

where $1 \le i \le n$, $1 \le j \le m$, and $I_n(i,j)$ denote the gray value of the pixels located at i columns and j rows in the image I_n, and T_1 is the threshold.

A novel pattern match-based adaptive background update method is expressed as follows. For an image sequence I, there is a classification result B; i.e., for any element $b_i \in B$, if $I_1, I_2 \in b_i$, then we have $I_1, I_2 \in I$ and $I_1 \sim I_2$. b_i has such properties as

Fig. 2.5 The Result Background of Selection Average Method (T1 = 50 and T2 = 3)[(a)] (a) The initial background is captured by multi-frame average method proposed in next section

Fig. 2.6 The Result Background of Selection Average Method (T1 = 50 and T2 = 3)[(a)] (a) The initial background is captured by pattern classification-based method proposed in next section

- $b_i.img = \frac{\Sigma_{I \in b_i} I}{count(B)}$;
- $b_i.lasttime = X$, where X is the maximum suffix of the element in b_i; and
- $b_i.fitness = X - Y$, where X is the maximum suffix of the element in b_i and Y is the minimal one.

Now, the background is defined as $b_i.img$, where $b_i.fitness > Th_f$ *and* $b_i.lasttime > Time_{current} - Th_t$. Here, Th_f and Th_t are two thresholds. There may be many b_i that satisfy this restriction. We can simply choose the one that fits best, or we can deem all of them to be possible background. This method is rewritten into the algorithm below.

Step 1: Set $\mathbb{B} = \emptyset$ as a set of candidate background. Here, $b \in \mathbb{B}$ has three fields: $b.img$, $b.lasttime$ and $b.fitness$, which are the background image, the time when $b.img$ was captured, and the fitness value of this background candidate.

Step 2: Capture a image $I_{current}$ from the camera, record the time $t_{current}$ from the real-time clock.

Step 3:

$$
\begin{aligned}
&if(I_{current} \sim b \text{ and } b \in \mathbb{B})\{\\
&\qquad b.img = k * I_{current} + (1-k) * b.img;\\
&\qquad b.fitness = b.fitness + (t_{current} - b.lasttime);\\
&\qquad b.lasttime = t_{current};\\
&\qquad goto\ \textbf{Step 2};\\
&\}
\end{aligned}
\tag{2.8}
$$

Step 4:

$$
\begin{aligned}
&if(I_{current} \sim I_{current-1})\{\\
&\qquad allocate\ a\ new\ candidate\ b_{current}.\\
&\qquad b_{current}.img = k * I_{current} + (1-k) * I_{current-1};\\
&\qquad b_{current}.fitness = 0;\\
&\qquad b_{current}.lasttime = t_{current};\\
&\qquad add\ b_{current}\ into\ \mathbb{B};\\
&\}
\end{aligned}
$$

Step 5:

$$
\begin{aligned}
&if(element\ number\ in\ \mathbb{B} > POLLMAX)\\
&\quad delete\ a\ candidate\ b_0\ in\ \mathbb{B},\ which\ has\ the\ lowest\ fitness,\ and\ is'nt\ b_{current};
\end{aligned}
$$

Step 6:

$$
\begin{aligned}
&for\ any\ b \in \mathbb{B}\\
&if(b.fitness > Th_f\ and\ b == b_{current});\\
&\quad b_{active} = b;\\
&else\\
&\quad set\ b_{active}\ as\ the\ one\ has\ the\ largest\ fitness.;
\end{aligned}
$$

Step 7: Repeat **Step 6** again to get b_{active_2}.
Step 8: Output b_{active} and b_{active_2};
Step 9: Goto **Step 2**;

Th_f is a threshold of time, which indicates when the background should be updated after it changes. *POLLMAX* is set to 32. In the results, b_{active} is the updated background, and b_{active_2} is a candidate background, which should be the background in shadow.

Figure 2.7 shows selected frame segments in the image sequence, which are 80×80 pixels in size. In the figure, the number below the frames is the frame ID, and can is the variable *lasttime* in Equation 2.8. Figure 2.8 shows the classes taken from the frame segments. In this case, the background should be the class $\{301\}$, whose *fitness* is 297. The results for this method run with the same video stream are illustrated in Figure 2.9, which shows that all moving objects are eliminated. Figure 2.10 compares the subtraction results using the pattern classification- and average-based methods, which reveal that the former has the advantage over the latter.

The pattern classification method can easily be rewritten for color images. Figure 2.11 shows the color image after background subtraction.

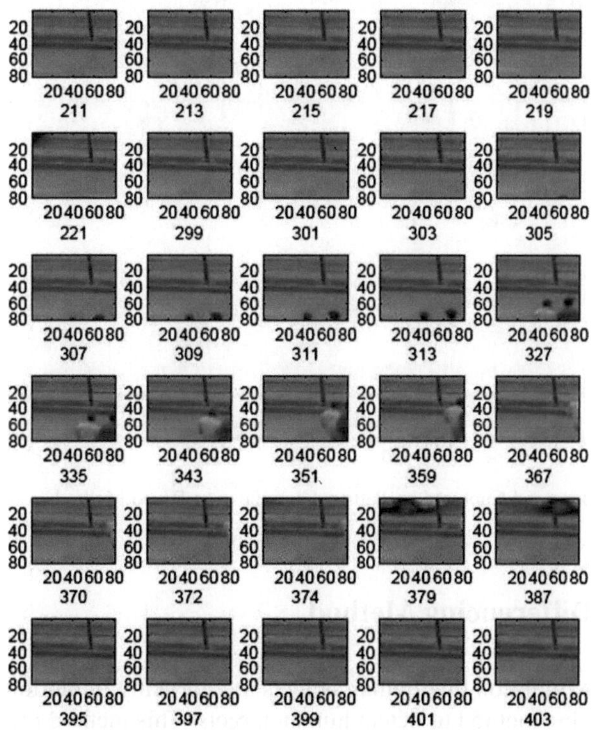

Fig. 2.7 A Selected Frame Segments

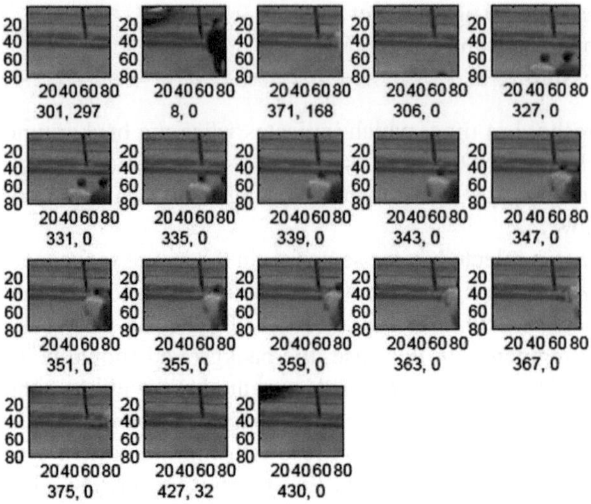

Fig. 2.8 The Classes in The Segments

Fig. 2.9 The Background Captured by Pattern Classification Based Method

2.3 Frame Differencing Method

Based on the assumption that human objects are normally in motion, we adopt the frame differencing method to detect human targets. This method requires only that the camera be kept static for a certain period, so it is applicable to moving cameras.

Frame differencing is the simplest method for the detection of moving objects, because the background model is equal to the previous frame. After performing a binarization process with a predefined threshold using the differencing method, we can find the target contour, and the target blob is obtained through the contour-filling process. However, if the target is moving very quickly, then the blob may contain too many background pixels, whereas if the target is moving very slowly, then target information in the blob may be lost. It is impossible to obtain solely foreground pixels when using the frame difference as the background model, but by using the following method, we can remove the background pixels and retrieve more foreground pixels, on the condition that the color of the foreground is not similar to that of the pixels of the nearby background. By separately segmenting the foreground and background in a rectangular area, we can label and cluster the image in the rectangular area again to obtain a more accurate foreground blob.

Figure 2.12 shows the foreground detection process in the frame differencing method, where (a) and (b) show the target detection results using the frame differencing method and blob filling results, respectively. In Figure 2.12 (c), it can be seen that the left leg of the target is lost. After the labeling and clustering process, we can retrieve the left leg (see Figure 2.12 (d)).

Feature selection is very important in tracking applications. Good features result in excellent performance, whereas poor ones restrict the ability of a system to distinguish the target from the background in the feature space. In general, the most desirable property of a visual feature is its uniqueness in the environment. Feature selection is closely related to object representation, in which the object edge or shape feature is used as the feature for contour-based representation, and color is used as a feature for histogram-based appearance representation. Some tracking algorithms use a combination of these features. In this chapter, we use color and spatial information for feature selection. The apparent color of an object is influenced primarily by the spectral power distribution of the illumination and the surface reflectance properties of the object. The choice of color and space also influences the tracking process. Digital images are usually represented in the red, green, blue (RGB) color space. However, the RGB color space is not a perceptually uniform color space because the differences among the colors do not correspond to the color differences perceived by humans. Additionally, RGB dimensions are highly correlated. Hue, saturation, and value (HSV) give an approximately uniform color space, which is similar to that perceived by humans; hence, we select this color space for this research. The color information alone is not sufficient. If we combine it with the spatial distribution information, then the selected features become more discriminative.

The main task is to segment the image using this feature. We choose the spatial-color mixture of Gaussians (SMOG) method to model the appearance of an object and define the Mahalanobis distance and similarity measure [168]. We then employ the k-means algorithm followed by a standard expectation maximization (EM) algorithm to cluster the pixels. Our approach is different in that we do not cluster and track the whole region but only the moving target in a rectangle, as described in the

Fig. 2.10 Example of Background Subtraction Result Operated By: (a) Pattern Classification Method, (b) Average Method

previous section. Figure 2.13 shows the clustering and tracking results for the whole region of the rectangle, and Figure 2.14 shows them for the moving target.

A standard method that clusters and tracks objects in the whole region of a rectangle can track targets properly but requires more particles and computation time, because the whole region contains many background pixels. When we choose a new particle at any place in the image, the similarity coefficient is likely to be high. Thus, more particles are required to find good candidate particles from the complex

Fig. 2.11 The Color Version of The Background Subtraction Result

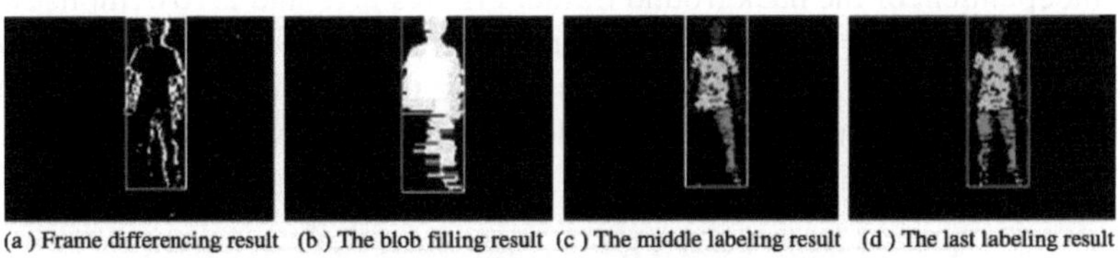

Fig. 2.12 Foreground detection process by frame differencing

background. We use the standard method to track a target in a real case, and find that the frame rate can reach 10 frames per second (fps) at a resolution of 160*120 pixels.

In contrast, if we cluster and track only the moving target, then fewer particles and less time are needed and the frame rate can reach 15 fps. To save computation time, therefore, the system clusters and tracks only moving targets.

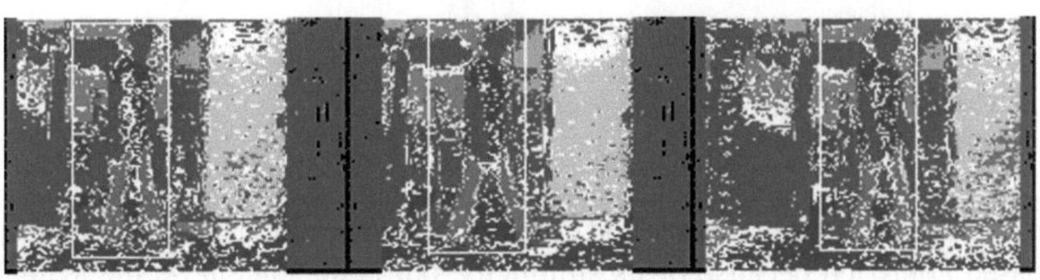

Fig. 2.13 Clustering and tracking results on the whole region in rectangles

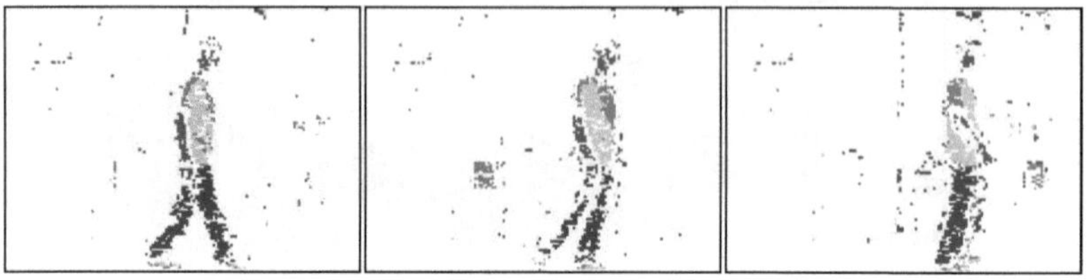

Fig. 2.14 Clustering and tracking results on moving targets

2.4 Optical Flow Method

Blob detection is the first task in video surveillance systems, because the precise process of segmentation and tracking depends heavily on the ability of a system to distinguish the foreground from the background. The usual approach is to subtract the background image, but such an approach fails when the environment becomes crowded as the background image becomes unavailable. Optical flow is one way to solve this problem [51] [57]. In our research, we use an optical flow algorithm that is independent of the background frame. Figures 2.15 and 2.16 demonstrate the experimental results for blob detection using optical flow. In a crowded environment in which a background frame is not available, the traditional subtraction algorithm fails to detect blobs, whereas the optical flow one performs well.

Fig. 2.15 Current frame

The traditional subtraction method can process information faster than the optical flow-based approach because of its simplicity, but is dependent on the availability of a background image. This means that in crowded environments, including campuses, shopping malls, subways, and train stations, the subtraction approach will fail, whereas the optical flow-based approach will work well. Although the latter approach is expensive in terms of computation, the use of super workstations makes its application feasible.

2.5 Conclusion

In this chapter, we introduce the first processing step, which is based on the acquired image, i.e., extracting the foreground, which the latter steps concern, from the background. Background and foreground detection are two sides of the same problem, which means that the solution of one will result in the solution of the other.

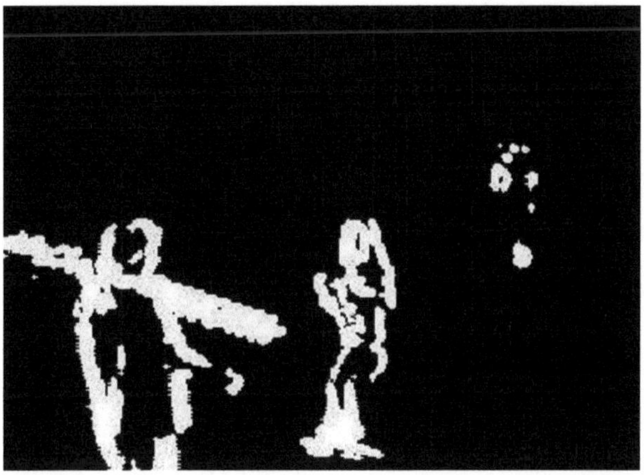

Fig. 2.16 Blob detection by optical flow

One approach is to develop a background model based on the pattern classification method, which adaptively updates the background, even at traffic crossings with heavy vehicle and pedestrian streams. Another is to identify moving targets in the foreground based on dynamic image information. The frame differencing method utilizes two consecutive frames to detect moving contours followed by a filling process to obtain the foreground. The optical flow method to detect moving objects is another foreground detection approach.

After the detection of the background and foreground, we can analyze the foreground blobs and connect different blobs in the spatial-temporal relationship for a given time domain.

Chapter 3
Segmentation and Tracking[1,2]

3.1 Introduction

After foreground extraction, if nonsignificant merging occurs among people, then segmentation is necessary, and the merged blobs can be split into separate ones equal in number to the number of people in the merged blob. Thereafter, all of the blobs can be indexed and tracked for motion and behavior analysis.

3.2 Segmentation

The position and orientation of a camera influence the degree of occlusion. If the camera's optical axis is horizontal, then occlusion will occur in the segment parallel to the horizontal axis of the image plane. Occluded blobs will be indistinguishable due to overlapping in depth.

To increase the segmentation capability from being one dimensional (1D) (horizontal axis of the image) to two dimensional (2D) (the horizontal and vertical axes of the image plane), the camera needs to be installed relatively high and tilted down for inspection.

We develop a hybrid segmentation methodology to split occluded blobs, using a histogram-based approach for horizontal occlusion and an ellipse-based approach for vertical occlusion.

---

[1] Portions reprinted, with permission, from Huihuan Qian, Xinyu Wu, Yongsheng Ou, and Yangsheng Xu, Hybrid Algorithm for Segmentation and Tracking in Surveillance, *Proceedings of the 2008 IEEE International Conference on Robotics and Biomimetics*. ©[2009] IEEE.

[2] Portions reprinted, with permission, from Zhixu Zhao, Shiqi Yu, Xinyu Wu, Congling Wang, and Yangsheng Xu, A multi-target tracking algorithm using texture for real-time surveillance, *Proceedings of the 2008 IEEE International Conference on Robotics and Biomimetics*. ©[2009] IEEE.

H. Qian et al., *Intelligent Surveillance Systems*, Intelligent Systems, Control and Automation: Science and Engineering 51, DOI 10.1007/978-94-007-1137-2_3,

To measure successful segmentation, evaluation rules must first be established. We introduce the following three rules:

- R1) The number of blobs in the segmentation results should be the same as the number of persons in that blob;
- R2) The areas of the segmented blobs in the histogram-based approach should be reasonably large; and
- R3) The shape of each blob after segmentation should be the filling rate of each ellipse in the ellipse-based approach, which should be larger than a certain threshold.

Before segmentation, it is necessary to filter out blobs of individuals, which do not need to be segmented, to reduce the computation cost and enhance real-time performance. The area threshold of a blob is not efficient because the area varies significantly, even though it contains only a single person, if the person makes great changes in posture or moves into different areas (depths) of the image.

We assume that in a less crowded environment, when entering the camera's field of view, a person first appears as an individual blob. When blobs merge, we can compute the number of persons in the occluded blob according to the take-up rate of its circum rectangle by the number of individual blobs in the previous frame, as illustrated in Figure 3.1.

Fig. 3.1 Human counting

The blue curves are the contours of two separate persons in frame number k, and the green rectangles are their corresponding circum rectangles. In frame number $k+1$, they are occluded, and the black rectangle is the circum rectangle of that occluded blob. The occluded blob's area is overlapped by the two separate rectangles in the previous frame at a percentage of about 30% and 52%, respectively, and thus should be segmented into two blobs; otherwise, the segmentation will have failed according to R1.

Two approaches are employed sequentially in the segmentation process: 1) the histogram-based approach, which depends on the peak and valley features in the vertical projection histogram [162]; and 2) the ellipse-based approach, which uses an ellipse to cut off merging persons from the top to the bottom [164].

(a) occlusion blob (b) segmentation result.

Fig. 3.2 Example of histogram-based segmentation.

Blobs, which become slightly occluded at the same depth, such as those in Figure 3.2(a), usually have special shape-based features. When projected vertically onto a horizontal line, the blobs change into the shapes shown in Figure 3.3. The curve is similar to a signal in the time domain, where the horizontal line is the time axis, and thus discrete Fourier transformation (DFT) can be employed to transform the line into a frequency domain. Thereafter, the 20 lowest frequency components are retained and the rest are set to zero by a filter. Using inverse discrete Fourier transformation (IDFT), they are then transformed back into the time domain. This process smoothes the original curve and benefits the following segmentation. On the smooth curve, the significant peaks are usually located at the horizontal position of the head, whereas the significant valleys are usually located at the effective cut points for segmenting the occlusion. An effective segmentation is shown in Figure 3.2(b).

Histogram-based segmentation fails in cases where people are occluded at different depths, a problem that also challenges W^4. To ensure that such false segmentation is avoided, we propose R2 to differentiate unsuccessful segmentation if the results contain blobs that are too small, as illustrated in Figure 3.4.

The histogram-based approach also fails if the blob cannot be segmented because the smoothed histogram curve does not bear two peaks, as shown in Figure 3.5.

The main shortcoming of the histogram-based approach is occlusion at different depths, which can be overcome through the integration of the approach with the ellipse-based one.

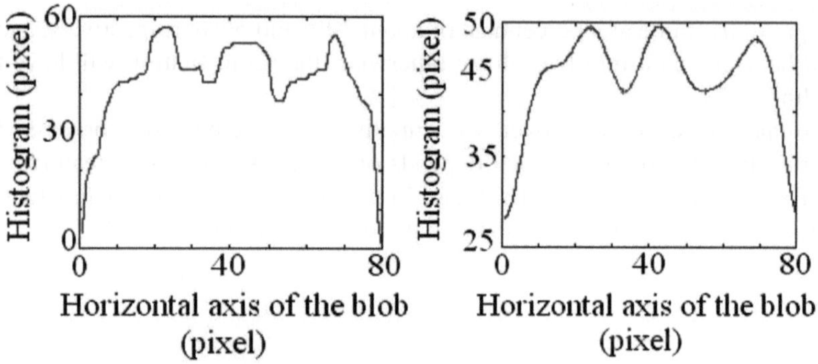

(a) vertical projected histogram. (b) smooth histogram after DFT and IDFT.

Fig. 3.3 Histogram curves. The red points correspond to the peaks that are the heads, and the valleys that are the effective cuts.

Fig. 3.4 Error in the segmentation result without R2.

In the ellipse-based approach, every person is assigned an ellipse, the long axis of which is aligned vertically to fit the normal walking shape. For simplicity, the size of the ellipse is the same, but it can be adjusted experimentally according to depth information. Given the number of persons (i.e., n) contained in the occlusion blob, we can split the vertically merging blob into n parts. In the first step, the person whose head is at the top of the occlusion blob is cut off using the ellipse by matching the top points of the blob and the ellipse. The small isolated blob fragments that result from the cut are treated as noise and filtered out. The remaining $n-1$ persons are split sequentially thereafter, as shown in Figure 3.6.

The validation of successful segmentation using the ellipse-based approach depends on the filling rate by the body in the ellipse. A threshold is chosen experimentally to cancel unsuccessful segmentation.

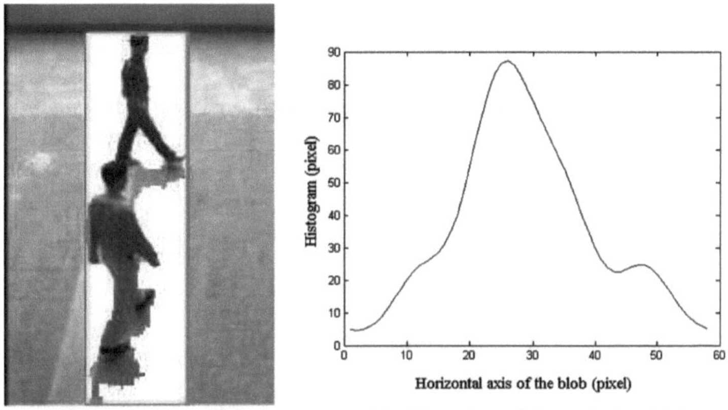

(a) Two-person's occlusion. (b) The histogram of the occlusion blob.

Fig. 3.5 Example of the failure of the histogram-based approach.

Fig. 3.6 Successful segmentation by using an ellipse-based algorithm.

The ellipse-based approach is not as stable as the peak valley one, primarily because the location of the topmost point of the ellipse is subject to the topmost pixel of the blob or remaining blob, and the lack of a filtering process makes it sensitive to noise. In addition, its validation method sometimes fails. Hence, this approach is inferior to the histogram-based one in the ranking of segmentation approaches. The flow chart for segmentation is illustrated in Figure 3.7.

3.3 Tracking

The objective of a surveillance system is to continuously look for suspicious people. In our research, we consider three cases: non-occlusion, segmentable occlusion, and unsegmentable occlusion. A hybrid approach is developed, which combines distance and color tracking.

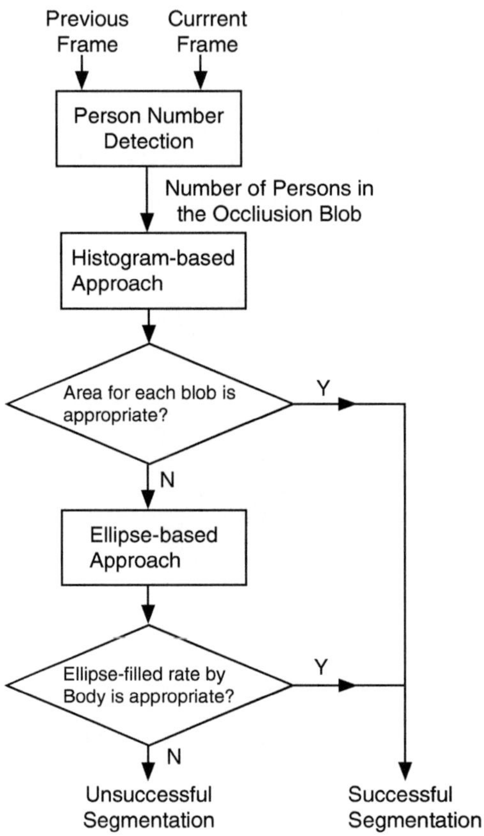

Fig. 3.7 Segmentation flow chart

3.3.1 Hybrid Tracking Method

3.3.1.1 Distance Tracking

To effectively track persons, we must first know their position. Therefore, it is important to choose the candidate point of their location. In the video experiment, the topmost pixel point of an individual blob is not a suitable choice because of the extreme influence of noise, which is disastrous for the rapid estimation of the one-order speed model. However, if we scan the line that is a few pixels down from the top one, then the method is less sensitive to noise and more robust. The position of an individual person is thus represented by this semi-topmost point.

In tracking, each blob is given an index number. In the representation previously defined, blobs are matched by the matching of the semi-topmost points. The distances between the points in the previous point set, which has been indexed, and those of the current point set can be investigated to determine the index of the current points.

However, with the information on points in the previous frames, a prediction model can be extracted, which will make the matching more accurate. A simple one-order speed model (1) can be established, which is both effective and computationally economical.

$$\hat{P}(n+1) = 2P(n) - P(n-1), \tag{3.1}$$

where $P(n)$ is a $2 \times N$ matrix with each row representing the 2D position of one semi-topmost point in the nth frame, $\hat{P}(n+1)$ is the prediction of $P(n+1)$, and N is the total number of people in the scene. Each un-indexed point in the set of the current frame is then matched to the closest indexed point in the predicted set.

3.3.1.2 Color Tracking

Distance tracking is effective when occlusion is not severe and the tracked targets do not suddenly change direction. However, when targets purposely change direction to fool the system, mistakes are made. Color tracking can be introduced to compensate for this deficiency. In color tracking, an individual blob has another representation parameter, i.e., the color spectrum, which describes the distribution of pixel colors in a blob in the hue space. The procedure is described as follows.

1. Convert the blob from RGB to HSV space.
2. Scan each pixel in the blob for its hue value, and span the blob into a histogram in which the horizontal axis represents the hue value and the vertical axis the number of pixels with the corresponding hue value, as illustrated in Figure 3.8.
3. Normalize the hue distribution by dividing each number by the total number of pixels in the blob.
4. Employ DFT to transform the histogram into a frequency domain.
5. Keep the low frequency components up to the twentieth one, and employ IDFT to transform the histogram back into a time hue domain.

The color spectrum results are shown in Figure 3.9, from which we can see that three different persons are represented by different color spectrums because of their clothes. The dashed, dashed-dotted, and solid curves each have one significant peak but at different locations, because persons A and C are apparently wearing the same

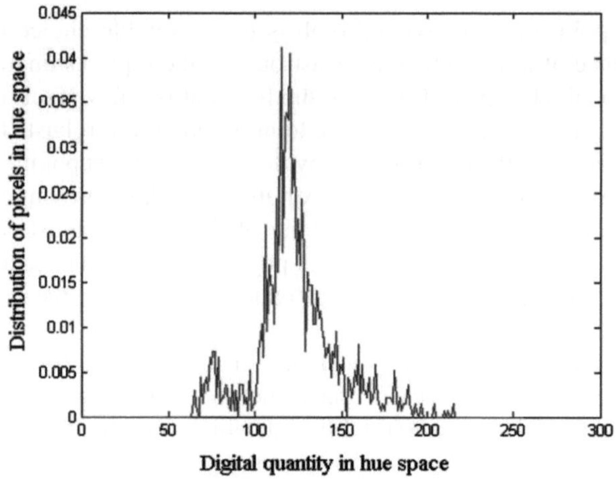

Fig. 3.8 Color spectrum of one person before DFT-IDFT.

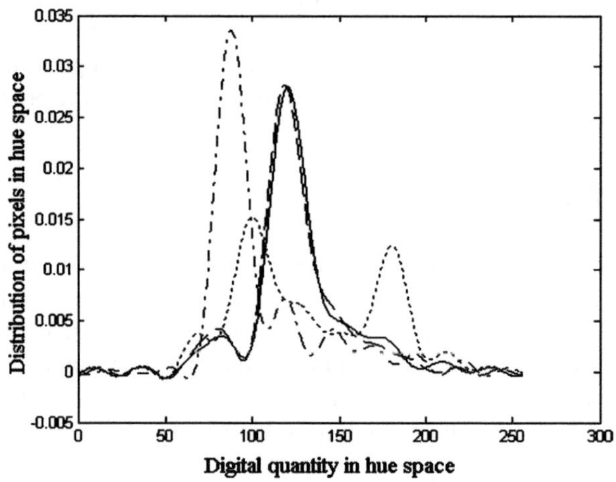

Fig. 3.9 Color spectrums of three persons for matching. The dashed, dotted, and dash-dotted curves correspond to the color spectrums of person A, B, and C in the previous frame, respectively. The solid curve is the color spectrum of person A in the current frame. The horizontal axis carries values from 0 to 255 that correspond to the 8 bits, and the vertical axis is the distribution in the color space.

color. The two peaks in the dashed curve show that the color of person B's clothes is less distinct compared to that of the others'. If the number of persons is small, then just two features of the spectrum, the mean and the variance, are sufficient for its successful application.

During video surveillance, people occasionally move out of the camera's field of view and re-enter it, which results in blobs being assigned another index number.

Color tracking can help to solve this problem. Instead of assigning a new index, the color tracking algorithm searches in its "memory" for a similar color distribution to make an optimal index assignment.

3.3.1.3 Fusion of the Two Tracking Approaches

Distance tracking and color tracking can be applied sequentially together with segmentation methods. Distance tracking is carried out continuously, whereas color tracking functions when distance tracking fails.

3.3.1.4 Experimental Study

In the experiment, three people walk in a 2D plane. They challenge the surveillance system by making unpredictable turns and joining together as one group.

The 2D movement space and large amount of background noise increase the difficulty for the individual segmentation algorithms. However, Figure 3.10 shows that when the histogram- and ellipse-based algorithms are both employed during the process, the results are successful.

Tracking both during and after the occurrence of an occlusion is also significantly difficult because of the unpredictable changes in the speed and direction of people. However, the hybrid tracking algorithm manages to correctly track each person, as shown in Figure 3.10. Each person is labeled with one index. Only in seriously occluded situations, such as in (d), is the blob unsegmentable.

3.3.2 Particle Filter-based Tracking Method

We also conduct research into human object tracking using an active camera. Activeness enlarges the camera's vision field; however, it also prevents robust background acquisition. Hence, in this scenario, we use the frame differencing method to obtain the rough foreground, which results in a higher level of scattered background noise compared to the background subtraction approach.

Here, we do not need to keep the target in the exact center of the scene, because doing so would require frequent movement of the camera and make it difficult to accurately detect the speed of the target. As shown in Figure 3.11, when the target's center is located between the left and right edges, the camera and mobile platform remain static. When the target moves and its center reaches the left or the right edge, the robot moves to keep the target in the center of the scene, according to the predicted results. When the camera moves, a particle filtering algorithm is employed to perform the tracking because it can overcome the difficulty of changes in the background.

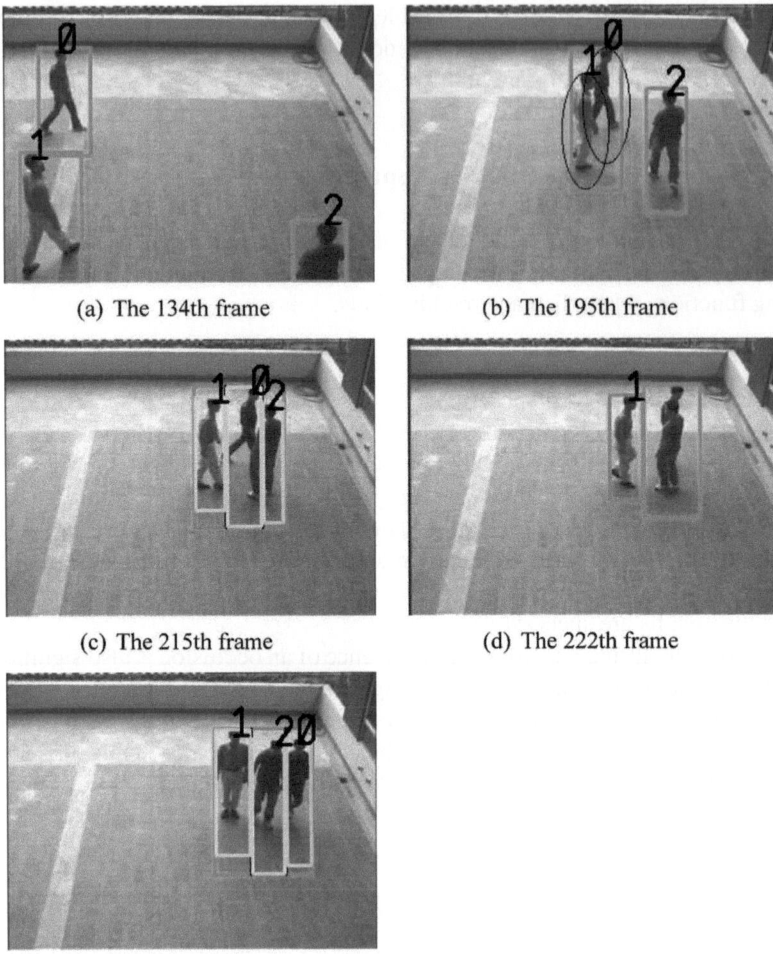

(a) The 134th frame

(b) The 195th frame

(c) The 215th frame

(d) The 222th frame

(e) The 251th frame

Fig. 3.10 Experimental result.

To tackle this problem, we adopt sequential Monte Carlo techniques, which are statistical methods and also known as particle filtering and condensation algorithms [169][170][171]. They have been widely applied in visual tracking in recent years. The general concept is that if the integrals required for a Bayesian recursive filter cannot be solved analytically, then the posterior probabilities can be represented by a set of randomly chosen weighted samples. The posterior state distribution $p(x_k|Z_k)$ needs to be calculated at each time step. In the Bayesian sequential estimation, the filter distribution can be computed by the following two-step recursion.

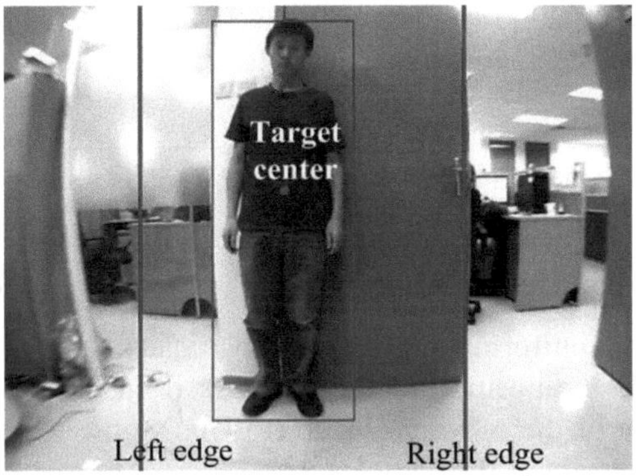

Fig. 3.11 The left edge and right edge for the tracking strategy.

Prediction step:

$$p(x_k|Z_{k-1}) = \int p(x_k|x_{k-1})p(x_{k-1}|Z_{k-1})dx_{k-1} \tag{3.2}$$

Filtering step:

$$p(x_k|Z_k) \propto p(z_k|x_k)p(x_k|Z_{k-1}) \tag{3.3}$$

Based on a weighted set of samples $\{x_{k-1}^{(i)}, \omega_k^{(i)}\}_{i=1}^N$ approximately distributed according to $p(x_{k-1}|Z_{k-1})$, we draw particles from a suitable proposed distribution, i.e., $x_k^{(i)} \sim q_p(x_k|x_{k-1}^{(i)}, z_k), i = 1, \ldots, N$. The weights of the new particles become

$$w_k^{(i)} \propto w_{k-1}^{(i)} \frac{p(z_k|x_k^{(i)})p(x_k^{(i)}|x_{k-1}^{(i)})}{q_p(x_k|x_{k-1}^{(i)}, z_k)} \tag{3.4}$$

The observation likelihood function $p(z_k|x_k)$ is important because it determines the weights of the particles and thus significantly influences tracking performance. We define the observation likelihood function using the SMOG method, which combines spatial layout and color information, as explained below.

Suppose that the template is segmented into k clusters, as described in the previous section. We then calculate the histograms, vertically and horizontally, of each cluster, and make a new histogram of $\tilde{\mathbf{q}}^i$ by concatenating the normalized vertical and horizontal histograms.

Regarding particles, pixels are first classified into k clusters on top of the template. Then, we calculate the normalized histograms of $\{\mathbf{q}_t^i(x_t)\}$, according to the template.

Following [181], we employ the following likelihood function for $p(z_k|x_k)$:

$$p(z_k|x_k) \propto \prod_i^k \exp -\lambda D^2[\tilde{\mathbf{q}}^i, \mathbf{q}_t^i(x_t)],$$

where λ is fixed to be 20 [181], and D is the Bhattacharyya similarity coefficient between two normalized histograms $\tilde{\mathbf{q}}^i$ and $\mathbf{q}_t^i(x_t)$:

$$D[\tilde{\mathbf{q}}^i, \mathbf{q}_t^i(x_t)] = \left[1 - \sum_k \sqrt{\tilde{\mathbf{q}}^i(k)\mathbf{q}_t^i(k; x_t)}\right]^{\frac{1}{2}}$$

The steps in the particle sample and updating process are as follows.

Step 1: Initialization: uniformly draw a set of particles.
Step 2: (a) Sample the position of the particles from the proposed distribution. (b) Find the feature of the moving object. (c) Update the weight of the particles. (d) Normalize the weight of the particles.
Step 3: Output the mean position of the particles that can be used to approximate the posterior distribution.
Step 4: Resample the particles randomly to obtain independent and identically distributed random particles.
Step 5: Go to the sampling step.
Computationally, the crux of the proportional fairness (PF) algorithm lies in the calculation of the likelihood.

3.3.2.1 Target Model Update

The target model obtained in the initialization process cannot be used throughout the whole tracking process because of changes in lighting, background environment, and target gestures. We therefore need to update the tracking target model in a timely way. However, if we update the target model at an improper time (e.g., when the camera is moving and the image is not clear), then the tracking will fail. Figure 3.12 shows a new target model updating process. In our camera control strategy, the camera remains static when the target is in the center of the camera view. When the camera is static, the frame differencing method is employed to obtain the target blob, and the similarity between the current and the initial blob is calculated. If the similarity property is larger than a given threshold, then we update the target model; otherwise, we move on to the frame differencing step. How to choose the threshold is an interesting problem. If the threshold is very large, then the similarity coefficient could easily be lower than the threshold, which will result in continual updating and consume much computation time. If the threshold is very low, then false detection is liable to occur. To balance the computation time and tracking results, we choose a threshold of 0.6 based on many experiments in real environments.

Figure 3.13 shows the unsuccessful tracking results due to not updating in an environment with changes in light. Figure 3.14 shows robust tracking results obtained using our update process in the same environment.

3.3.3 Local Binary Pattern-based Tracking Method

The local binary pattern (LBP) operator is a powerful means for describing texture. It is defined as an illumination intensity invariant texture measure, derived from a general definition of texture in a local neighborhood. There is no float point

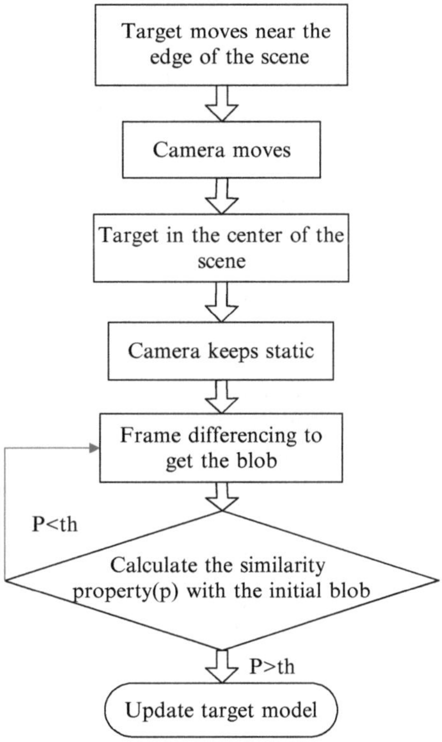

Fig. 3.12 Target model update process.

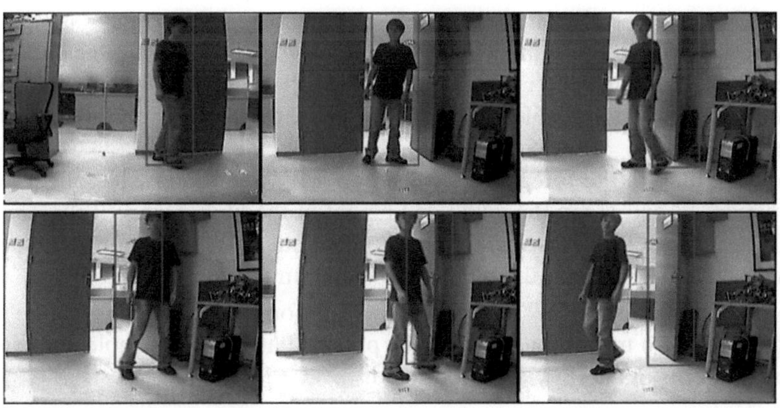

Fig. 3.13 Tracking results without update.

Fig. 3.14 Tracking results employing our update process.

operation in LBP feature extraction, and an LBP-based method can easily be implemented on a digital signal processor (DSP) platform. In a (P,R) neighborhood, a binary code that describes the local texture pattern is built by taking the threshold of the gray value from its center. The binary code, i.e., the LBP operator, is denoted as $LBP_{(P,R)}$. The notion (P,R) means that there are P sampling points on a circle of radius R. The LBP neighborhood can be divided into two groups, the square and circle neighborhoods, as shown in Figures 3.15 (a) and (b). Figure 3.16 (c) illustrates how the LBP operator is computed in the $(8,1)$ square neighborhood. The binary code of this example is 11010010_b. Thus, the LBP operator at g_c can be defined as

$$LBP_{(P,R)} = \sum_{p=0}^{P-1} s(g_p - g_c) \cdot 2^p \tag{3.5}$$

where $s(x) = \begin{cases} 1 \text{ if} x \geq 0 \\ 0 \text{ otherwise} \end{cases}$.

Uniform LBPs, which are an extension of basic LBPs, can describe the fundamental properties of local image textures [185]. A pattern is called uniform if there is at most one bitwise transition from 0 to 1 in the binary patterns, such as 00000000_b, 00011110_b. Ojala et al. found that uniform patterns comprise the vast majority, nearly 90% in the $(8,1)$ neighborhood and around 70% in the $(16,2)$ neighborhood, of the texture patterns in their experiments [185]. Nonuniform patterns, in contrast, provide statistically unreliable information. Hence, we use only uniform patterns in our algorithm.

For each blob in the foreground, we set up a $W \times H$ window to collect texture information. W and H are the width and height of the blob, respectively. The LBP code for each pixel in the window is computed, but only uniform patterns are reserved for the histogram of the blob. The histogram is denoted as $LBP^{u}_{(P,R)}2$. In the (P,R) neighborhood, if (P,R) is known, then uniform pattern codes can be determined, which also means that they can be precalculated. For example, there are 59 uniform patterns in the $(8,1)$ neighborhood.

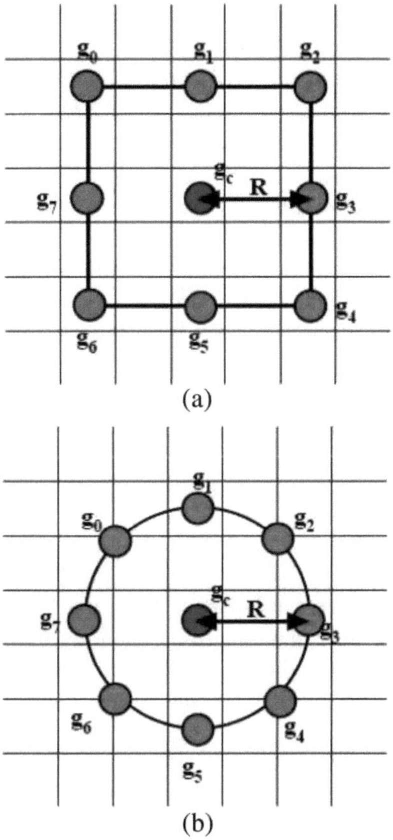

Fig. 3.15 Neighborhood

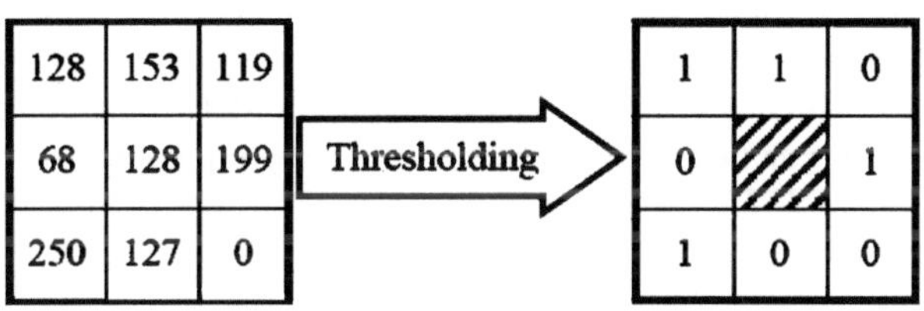

Fig. 3.16 LBP operator computing in a (8,1) square neighborhood

The texture histogram $H_{1\times N}$ can be defined as

$$H_i = \sum_{x=0}^{W-1} \sum_{y=0}^{H-1} \eta(LBP_{(P,R)}(x,y), i) \tag{3.6}$$

where $i = 0, \cdots, N-1$, N is the number of the patterns, and $\eta(L,i) = 1$ only if $L = i$. Among the 256 bins, only the 59 bins for uniform patterns are reserved to describe an object.

3.3.3.1 Multiple Target Tracking

Our objective is to implement the algorithm on a TI DM642 DSP platform, the limited computational resources of which, compared with a standard PC, pose great challenges for real-time tracking. Thus, to reduce computational complexity, we choose the single Gaussian method for background modeling and foreground segmentation instead of the mixture of Gaussians (MoG) one, even though the latter is better at handling leaf-shaking and illumination changes. Morphology operations are applied to remove isolated small spots and fill holes in the foreground. After these steps, we get some isolated blobs in the foreground of every frame. Undesirable blobs are deleted by a size filter. Each blob is characterized by the following features:

1) $\overrightarrow{C} = [X_c, Y_c]^T$, centroid of the blob;
2) blob width W and height H; and
3) uniform LBP histogram H_{LBP}.

Therefore, the ith blob in the mth frame can be represented as

$$B_i^m = \{C_i^m, W_i^m, H_i^m, H_{LBP_i}^m\} \tag{3.7}$$

The whole tracking procedure can be seen in Figure 3.17. First, a Kalman filter is applied to predict each blob's new position. Then, collision detection is executed. If just one candidate is found in the neighborhood, then the candidate is considered to be the tracking target. If more than one unoccluded blob is detected, then the LBP histogram is chosen as the feature to distinguish blobs.

3.3.3.2 Kalman Filter

The Kalman filter is an efficient recursive filter that can estimate the state of a dynamic system from a series of incomplete and noisy measurements [186, 187]. Suppose that y_t represents the output data or measurements in a system, the corresponding state of which is referred to as x_t. Then, the state transition from $t-1$ to t in the Kalman filter can be described by the following equation:

$$x_t = Ax_{t-1} + \omega_{t-1} \tag{3.8}$$

Its measurement y_t is

$$y_t = Hx_t + v_t \tag{3.9}$$

A is termed *state transition matrix*, and H is *measurement matrix*. ω_t and v_t are process and measurement noise, respectively, which are assumed to be independent, white, and normally distributed in the tracking.

In our system, a blob's $X-Y$ coordinate position (x,y), size (w,y), and position changes (dx,dy) are used as states, which can be represented by a vector $[x,y,w,h,dx,dy]^T$. The measurement vector is $[x,y,w,h]^T$. Thus, the state transition matrix is

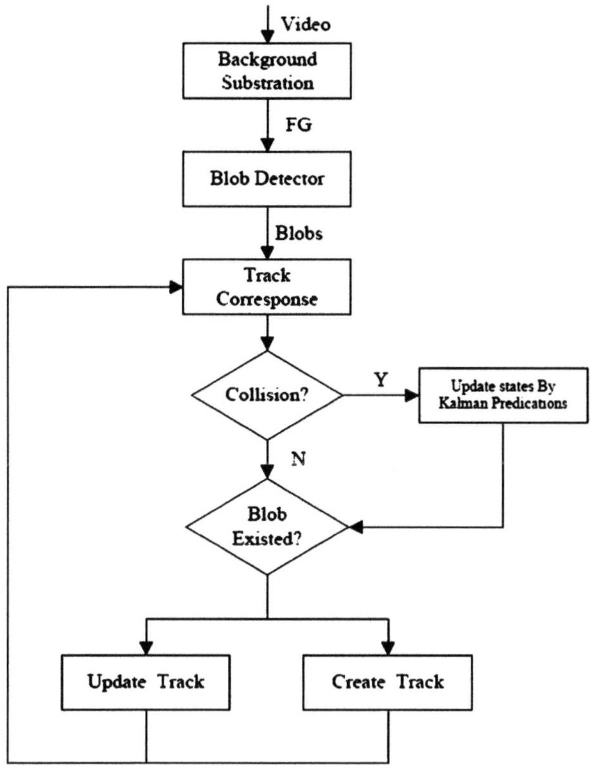

Fig. 3.17 Flowchart of the tracking system

$$A = \begin{bmatrix} 1\,0\,0\,0\,1\,0 \\ 0\,1\,0\,0\,0\,1 \\ 0\,0\,1\,0\,0\,0 \\ 0\,0\,0\,1\,0\,0 \\ 0\,0\,0\,0\,1\,0 \\ 0\,0\,0\,0\,0\,1 \end{bmatrix}$$

$$H = \begin{bmatrix} 1\,0\,0\,0\,0\,0 \\ 0\,1\,0\,0\,0\,0 \\ 0\,0\,1\,0\,0\,0 \\ 0\,0\,0\,1\,0\,0 \end{bmatrix}$$

Two steps are completed in the Kalman filter process: time and measurement updating. The time update step is in charge of prediction, and the measurement update step is responsible for feedback or correction. In normal tracking, a predicted blob is used for target searching. In the case of a collision, the blob's position and size properties are updated using Kalman predictions. We note that the Kalman filter is efficient in handling temporary collision problems.

3.3.3.3 LBP Histogram Distance

The LBP histogram is used to describe blobs, and a N-length texture histogram is created for each blob. Many kinds of measurements can be used to measure the similarity between blobs. The most commonly used are correlation, intersection, and the chi-square distance. These measurements are defined as follows:

- Correlation:

$$d(H_1,H_2) = sum_i \frac{H_1'(i) \cdot H_2'(i)}{\sqrt{(\Sigma_j H_1'(j)^2) \cdot (\Sigma_j H_2'(j)^2)}} \tag{3.10}$$

where $H_k'(j) = H_k(j) - \frac{\Sigma_j H_k(j)}{N}$

- Intersection:

$$d(H_1,H_2) = \sum_i \min(H_1(i), H_2(i)) \tag{3.11}$$

- Chi-square (χ^2) distance:

$$\chi^2(H_1,H_2) = \sum_{i=0}^{N-1} \frac{H_1(i) - H_2(i)}{H_i(i) + H_2(i)} \tag{3.12}$$

Among these measures, the computational complexity of correlation, $O(N^2)$, is more time consuming that that of the others, $O(N)$. According to Ahonen's experiments, the chi-square (χ^2) distance performs slightly better than histogram intersection [188]; thus, we use the chi-square distance in our experiments.

3.3.3.4 Blob Classification

Suppose that a blob B_i^t is detected in the tth frame and its Kalman prediction is $\overline{B}_i^t$, and we want to track it in the $(t+1)$th frame. The blobs in the $(t+1)$th frame are B_i^{t+1}, where $j = 1,2,\cdots,N$. The distance vector between blob $\overline{B}_i^{t+1}$ and blob B_i^{t+1} is

$$\overrightarrow{D}_j = \begin{bmatrix} Xc_{\overline{B}_i^t} - Xc_{\overline{B}_i^{t+1}} \\ Yc_{\overline{B}_i^t} - Yc_{\overline{B}_i^{t+1}} \end{bmatrix} \tag{3.13}$$

If a blob is near the predicted blob $\overline{B}_i^{t+1}$ and $|Xc_{\overline{B}_i^t} - Xc_{\overline{B}_i^{t+1}}| < S \cdot \frac{W_{\overline{B}_i^t} + W_{\overline{B}_i^{t+1}}}{2}$ and $|Yc_{\overline{B}_i^t} - Yc_{\overline{B}_i^{t+1}}| < S \cdot \frac{H_{\overline{B}_i^t} + H_{\overline{B}_i^{t+1}}}{2}$, then the blob B_t^{t+1} is considered to be a candidate of B_i^t. All candidates are incorporated into a sequence. Parameter S is used to adjust the search window so that blobs that move quickly can also be found. If just one blob is in the candidate sequence, then the blob is the tracking target of B_i^t and its properties will be incorporated into the tracking sequence of blob B_i^t. If multiple blobs exist, then we choose the one that is the most similar to the blob. The similarity is measured by the chi-square (χ^2) distance.

3.3.3.5 Experiment and Discussion

The proposed algorithm has been implemented on PC and TI DSP platforms and can run at a frame rate of 25 fps. We also test our algorithm using the PETS 2001 dataset and outdoor videos captured from a fixed camera outside our lab. PETS datasets [189] are designed for testing visual surveillance algorithms, and widely used by many researchers. The PETS2001 dataset used in our experiments is multi-viewed (two cameras) and 768×576 in size, and the size of the outdoor series is 320×240. Details are given in the table below. Pedestrians, vehicles, and bicycles are in the test videos, and collisions occur in the PETS 2001 and outdoor datasets.

Table 3.1 Test Video Details

Series	Dataset	View	Frames
PETS 2001	Dataset1	Camera1	2689
		Camera2	2689
Outdoor	Outdoor1	Camera1	106451
	Outdoor2	Camera1	11391
	Outdoor3	Camera1	17045

Experiments using the PETS dataset show that the LBP distance can be employed to distinguish blobs. We collect each blob's texture histogram from each frame and calculate its LBP distance using the blob texture histograms that are collected in the next frame. The LBP neighborhood is $P = 8$, $R = 1$ in our experiment. To ensure that all of the collected data are valid, leaf-shaking and illumination variations that cause spurious blobs are abandoned, and incorrect tracks, such as those losing targets or confusing the ID of targets, are fixed manually. Fifteen blobs are detected in PETS dataset 1 camera 1. Except for incorrect foreground detection (blobs 0, 2, 11, and 12), 11 blobs are correctly detected. Among the 11 blobs, blobs 1 and 14 exchange IDs at frame 2245. Thus, nine blobs are correctly tracked. The same operation is applied to the tracking of camera 2. In most cases, the average LBP distance of different blobs is greater than that of the same blob.

The tracking results are illustrated in Figure 3.18. The tracking blobs are indicated by black rectangles in each frame, and the trajectory of each blob is indicated by a corresponding blob of the same color. The following scenes are found in the test videos: (1) moving in one direction without collision, (2) wandering, and (3) collision. The results also show that clear differences exist between the LBP distance of same blobs and that of different blobs. As a whole, the LBP distance of the former is far less than that of the latter. In our outdoor tracking experiments, the

LBP distance of about 85% of the same blobs is less than 70, while that of most of the different blobs is greater than 100.

Fig. 3.18 Tracking results

The Kalman filter, combined with the LBP distance, demonstrates satisfactory performance in dealing with collision or occlusion problems. If collision is detected, then the properties of blobs are updated using Kalman predictions. The LBP distance is used to pick out the most similar tracking target before blobs merge and split. Blobs are correctly predicted and most trajectories under collision or occlusion continue reliably in our system. Figure 3.18(b) presents an example. As shown, blobs 19 and 22 collide at frame 3936. They are properly tracked, and their IDs are correctly preserved using our algorithm.

Typical tracking failures are shown in Figure 3.18(c). They occur if the motion of blobs suddenly changes. These failures generally occur at the edge of the camera's field of view. In this experiment, the white car (blob 432) stops in frame 46164, when a silver car passes by. The unexpected change in motion leads to the track loss of the white car.

3.4 Conclusion

In this chapter, we analyze the foreground blobs detected in Chapter 2. For those blobs in which more than two persons are occluded, segmentation is carried out. The hybrid segmentation approach combines histogram- and ellipse-based algorithms.

For the temporal analysis of people's behavior, tracking is necessary to link the segmented blobs in the time domain. Three approaches that have been developed by our R&D group are elaborated: hybrid tracking based on distance and color information, particle filter-based tracking for active cameras, and local binary pattern-based tracking, which utilizes texture information.

Following segmentation and tracking, detailed analysis of human behavior can be conducted in the monitored region.

Chapter 4
Behavior Analysis of Individuals[1,2]

4.1 Introduction

After successful segmentation and tracking, the temporal-spatial information of each blob sequence can be indexed for further behavior analysis. In this chapter, we elaborate two kinds of approaches for the behavior analysis of individuals: learning-based and rule-based.

4.2 Learning-based Behavior Analysis

4.2.1 Contour-based Feature Analysis

4.2.1.1 Preprocessing

In this step, we collect consecutive blobs for a single person from which the contours of a human blob can be extracted. As can be seen in Figure 4.1, the centroid (x_c, y_c) of the human blob is determined by the following equations.

$$x_c = \frac{1}{N}\sum_{i=1}^{N} x_i, y_c = \frac{1}{N}\sum_{i=1}^{N} y_i \tag{4.1}$$

[1] Portions reprinted, with permission, from Xinyu Wu, Yongsheng Ou, Huihuan Qian, and Yangsheng Xu, A Detection System for Human Abnormal Behavior, *IEEE International Conference on Intelligent Robot Systems*. ©[2005] IEEE.

[2] Portions reprinted, with permission, from Yufeng Chen, Guoyuan Liang, Ka Keung Lee, and Yangsheng Xu, Abnormal Behavior Detection by Multi-SVM-Based Bayesian Network, *International Conference on Information Acquisition*. ©[2007] IEEE.

H. Qian et al., *Intelligent Surveillance Systems*, Intelligent Systems, Control and Automation: Science and Engineering 51, DOI 10.1007/978-94-007-1137-2_4,

where (x_c, y_c) is the average contour pixel position, (x_i, y_i) represent the points on the human blob contour, and there are a total of N points on the contour. The distance d_i from the centroid to the contour points is calculated by

$$d_i = \sqrt{(x_i - x_c)^2 - (y_i - y_c)^2} \tag{4.2}$$

The distances for the human blob are then transformed into coefficients by means of DFT. Twenty primary coefficients are selected from these coefficients and thus 80 coefficients are chosen from four consecutive blobs. We then perform PCA for feature extraction.

Fig. 4.1 Target preprocessing.

4.2.1.2 Supervised PCA for Feature Generation

Suppose that we have two sets of training samples: A and B. The number of training samples in each set is N. Φ_i represents each eigenvector produced by PCA, as illustrated in [54]. Each of the training samples, including positive and negative samples, can be projected onto an axis extended by the corresponding eigenvector. By analyzing the distribution of the projected $2N$ points, we can roughly select the eigenvectors that have more motion information. The following gives a detailed description of the process.

1. For a certain eigenvector Φ_i, compute its mapping result according to the two sets of training samples. The result can be described as $\lambda_{i,j}, (1 \leq i \leq M, 1 \leq j \leq 2N)$.
2. Train a classifier f_i using a simple method such as perception or another simple algorithm that can separate $\lambda_{i,j}$ into two groups, normal and abnormal behavior, with a minimum error $E(f_i)$.
3. If $E(f_i < \theta)$, then we delete this eigenvector from the original set of eigenvectors.

M is the number of eigenvectors, and $2N$ is the total number of training samples. θ is the predefined threshold.

It is possible to select too few or even no good eigenvectors in a single PCA process. We propose the following approach to solve this problem. We assume that the number of training samples, $2N$, is sufficiently large. We then randomly select training samples from the two sets. The number of selected training samples in each set is less than $N/2$. We then perform supervised PCA using these samples. By repeating this process, we can collect a number of good features. This approach is inspired by the bootstrap method, the main idea of which is to emphasize some good features by reassembling data, which allows the features to stand out more easily.

4.2.1.3 SVM classifiers

Our goal is to separate behavior into two classes, abnormal and normal, according to a group of features. Many types of learning algorithms can be used for binary classification problems, including SVMs , radial basis function networks (RBFNs), the nearest neighbor algorithm, and Fisher's linear discriminant, among others. We choose an SVM as our training algorithm because it has stronger theory interpretation and better generalization performance than the other approaches.

The SVM method is a new technique in the field of statistical learning theory [53], [55], [56], and can be considered a linear approach for high-dimensional feature spaces. Using kernels, all input data are mapped nonlinearly into a high-dimensional feature space. Separating hyperplanes are then constructed with maximum margins, which yield a nonlinear decision boundary in the input space. Using appropriate kernel functions, it is possible to compute the separating hyperplanes without explicitly mapping the feature space.

In this subsection, we briefly introduce the SVM method as a new framework for action classification. The basic idea is to map the data X into a high-dimensional feature space F via a nonlinear mapping Φ, and to conduct linear regression in this space.

$$f(\overline{x}) = (\omega \cdot \Phi(\overline{x})) + b, \Phi : R^N \to F, \omega \in F \tag{4.3}$$

where b is the threshold. Thus, linear regression in a high-dimensional (feature) space corresponds to nonlinear regression in the low-dimensional input space R^N. Note that the dot product in Equation (4.3) between ω and $\Phi(\overline{x})$ has to be computed in this high-dimensional space (which is usually intractable) if we are not able to use the kernel, which eventually leaves us with dot products that can be implicitly expressed in the low-dimensional input space R^N. As Φ is fixed, we determine ω from the data by minimizing the sum of the empirical risk $R_{emp}[f]$ and a complexity term $\|\omega\|^2$, which enforces flatness in the feature space.

$$\begin{aligned} R_{reg}[f] &= R_{emp}[f] + \lambda \|\omega\|^2 \\ &= \Sigma_{i=1}^{l} C(f(\overline{x}_i - y_i)) + \lambda \|\omega\|^2 \end{aligned} \tag{4.4}$$

where l denotes the sample size $(\overline{x}_1, \ldots, \overline{x}_l)$, $C(\cdot)$ is a loss function, and λ is a regularization constant. For a large set of loss functions, Equation (4.4) can be minimized by solving a quadratic programming problem, which is uniquely solvable. It is shown that the vector ω can be written in terms of the data points

$$\omega = \Sigma_{i=1}^{l} (\alpha_i - \alpha^*) \Phi(\overline{x}_i) \tag{4.5}$$

with α_i, α^* being the solution of the aforementioned quadratic programming problem. α_i and α^* have an intuitive interpretation as forces pushing and pulling the estimate $f(\overline{x}_i)$ towards y_i the measurements. Taking Equations (4.5) and (4.3) into account, we are able to rewrite the whole problem in terms of dot products in the low-dimensional input space as

$$f(\overline{x}) = \Sigma_{i=1}^{l}(\alpha_i - \alpha^*)(\Phi(\overline{x}_i)\Phi(\overline{x})) + b$$
$$= \Sigma_{i=1}^{l}(\alpha_i - \alpha^*)K(\overline{x}_i, \overline{x}) + b \qquad (4.6)$$

where α_i and α^* are Lagrangian multipliers, and $\overline{x}$ are support vectors.

Fig. 4.2 A running person

4.2.1.4 Experiments

Our approach differs from previous detection methods in that (1) it can detect many types of abnormal behavior using one method and the range of abnormal behavior can be changed if needed; and (2) rather than detecting body parts ([53]) on the contour, 20 primary components are extracted by DFT. In our system, we do not utilize the position and velocity of body parts for learning, because the precise position and velocity of body parts cannot always be obtained, which causes a high rate of false alarms. In the training process, for example, we put the blobs of normal behavior, such as standing and walking, into one class and the blobs of abnormal behavior,

(a) Normal behavior (b) Abnormal behavior

Fig. 4.3 A person bending down.

(a) Normal behavior (b) Abnormal behavior.

Fig. 4.4 A person carrying a bar

such as running, into another one. PCA is then employed to select features, and an SVM is applied to classify the behavior as running or normal behavior. In the same way, we obtain classifiers for bending down and carrying a bar. The three SVM classifiers are then hierarchically connected.

After the SVM training, we test the algorithm using a series of videos, as illustrated in Figure 4.2, which shows that a running person is detected while other people are walking nearby; Figure 4.3, which shows that a person bending down is detected; and Figure 4.4, which shows that a person carrying a bar in a crowd is detected.

The algorithm is tested using 625 samples (each sample contains four consecutive frames). The success rate for behavior classification is shown in Table 4.1.

Table 4.1 Results of the classification.

Behavior		Success rate
Human abnormal behavior	Running	82% (107 samples)
	Bending down	86%(113 samples)
	Carrying a bar	87%(108 samples)
Human normal behavior		97%(297 samples)

4.2.2 Motion-based Feature Analysis

4.2.2.1 Mean Shift-based Motion Feature Searching

Before a feature can be extracted, we need to decide where the most prominent one is. In the analysis of human behavior, information about motion is the most important. In this section, we use the mean shift method to search for the region that has a concentration of motion information.

People generally tend to devote greater attention to regions where more movement is occurring. This motion information can be defined as the difference between the current and the previous image. Information about the motion location and action type can be seen in Figure 4.5.

(a) Action of hitting (b) Motion information of hitting

(c) Action of double waving (d) Motion information of double waving

Fig. 4.5 Motion information of human action

The process of the basic continuously adaptive mean shift (CAMSHIFT), which was introduced by Bradski [190], is as follows.

1. Choose the initial location of the search window and a search window size.
2. Compute the mean location in the search window and store the zeroth moment.
3. Center the search window at the mean location computed in the previous step.
4. Repeat Steps 2 and 3 until convergence (or until the mean location moves less than a preset threshold).
5. Set the search window size equal to a function of the zeroth moment found in Step 2.
6. Repeat Steps 2-5 until convergence (or until the mean location moves less than a preset threshold).

In our problem formulation, attention should be paid to the window location and size searching steps. Because a single feature is insufficient for the analysis of human behavior, more features are retrieved based on the body's structure. The head and body size can be estimated according to the method explained in section II.A, and the region of interest can be defined as shown in Figure 4.6.

Fig. 4.6 Feature selection: The white rectangle is the body location detected and the four small colored rectangles are the context confined region for feature searching.

Here, four main feature regions are selected according to human movement characteristics, and more detailed regions can be adopted depending on the requirements. The locations of the search windows are restricted to the corresponding regions, and the initial size is set to be the size of a human head. For Step 2, the centroid of the moving part within the search window can be calculated from zero and first-order moments [191]: $M_{00} = \sum_x \sum_y I(x,y)$, $M_{10} = \sum_x \sum_y xI(x,y)$, and $M_{01} = \sum_x \sum_y yI(x,y)$, where $I(x,y)$ is the value of the difference image at the position (x,y).

The centroid is located at $x_c = \frac{M_{10}}{M_{00}}$ and $y_c = \frac{M_{01}}{M_{00}}$.

More information can be inferred from higher-order moments, such as direction and eigenvalues (major length and width). The general form of the moments can be written as $M_{ij} = \sum_x \sum_y x^i y^j I(x,y)$.

Then, the object orientation is $\theta = \frac{\arctan \frac{2b}{a-c}}{2}$, where $a = \frac{M_{20}}{M_{00}} - x_c^2$, $b = \frac{M_{11}}{M_{00}} - x_c y_c$, and $c = \frac{M_{02}}{M_{00}} - y_c^2$.

4.2.2.2 Motion History Image-based Analysis

The distribution of the difference image is limited to human motion understanding. For the recognition of action sequences, information about the direction of movement is more important than that of static poses. In this section, we introduce a real-time computer visual representation of human movement known as a motion history image (MHI) [192, 193], which can help generate the direction and the gradient from a series of foreground images.

The MHI method is based on the foreground image, and an appropriate threshold is needed to transform the image into a binary one. If the difference is less than the threshold, then it is set to zero. The first step is to update the MHI template T using the foreground at different time stamps τ:

$$T(x,y) = \begin{cases} \tau, & I(x,y) = 1 \\ 0, & T(x,y) < \tau - \delta \\ T(x,y), & else \end{cases}$$

where I is the difference image, and δ is the time window to be considered.

The second step is to find the gradient on this MHI template. $\theta(x,y) = \arctan \frac{G_y(x,y)}{G_x(x,y)}$, where $G_y(x,y), G_x(x,y)$ are the sobel convolution results with the MHI template in the x and y directions, respectively. The scale of the filter can vary in a large region according to the requirements. However, the gradient on the border of the foreground should not be considered, for the information is incomplete and using it may lead to errors. Then, the regional orientation of the MHI can be obtained from the weighted contribution of the gradient at each point, where the weight depends on the time stamp at that point.

4.2.2.3 Frame Work Analysis

Automatic recognition of human action is a challenge, as human actions are complex under different situations. Some simple actions can be defined statically for the recognition of basic poses, which is an efficient way to provide basic recognition of action types. However, if no temporal information is taken into consideration, then it is difficult to describe actions that are more complex. Some researchers have tried to take into account the temporal dimension of actions. Most of such work is centered on specific action sequences, such as the analysis of predefined gestures using the hidden Markov model (HMM) and dynamic programming. In the analysis of human actions, such models are very limited in terms of their effectiveness in the recognition of complex action sequences.

In maintaining the accuracy of a recognition system, there is a trade-off between the complexity of action and the range of action allowed. In contrast to typical gesture recognition applications, intelligent surveillance concentrates on detecting abnormal movements among all sorts of normal movements in the training data rather than accurately recognizing several complex action sequences.

It is not always easy to define what behaviors are abnormal. Generally, two approaches are used: one is to model all normal behaviors, so that any behavior that does not belong to the model will be recognized as abnormal, and the other is to directly model abnormal behaviors. As it is not practical to list all abnormal behaviors that may happen in a scene, we prefer the first approach. In this way, a limited number of typical normal behaviors can be added dynamically into the database one after another.

Given a feature space F and the feature evidence $X, X \subseteq F$, we try to find a normal model M to express the probability $P(\overline{H^0}|x)$ of hypothesis $\overline{H^0}$, where H^0 is defined as the case that the candidate feature $x (x \in X)$ belongs to abnormal behavior, and $\overline{H^0}$ belongs to normal behavior. In addition, types of action $H^i, i \in N$ can be recognized using different kinds of action models $M_i, i \in N$.

4.2.2.4 SVM-based Learning

In this part, to simplify the optimization procession, we take the radial basis function (RBF) as a kernel for Equation (4.6):

$$K(\overline{x}_i, \overline{x}) = \exp(-\gamma \|\overline{x}_i - \overline{x}\|^2), \gamma > 0 \tag{4.7}$$

where γ is the kernel parameter. Then, the optimized results can be found in the extended space by parameter γ and loss function $C(\cdot)$.

Once the function is obtained, the classification line and the probability of the data can be directly calculated. For each action type, we establish an SVM model to give the probability $P(x|H_i)$ that evidence $x, x \in X$ belongs to each hypothetical action type H_i.

4.2.2.5 Recognition using a Bayesian Network

Generally, the hypothesis that has the maximum probability among all of the models is selected as the recognized action type. If all of the probabilities are smaller than a certain threshold value, then the action is recognized as an abnormal behavior.

However, the recognition of actions directly based on the maximum probability of all of the hypotheses is limited, as some of the poses in different action sequences are very similar. Therefore, the relationship of the context frame should be incorporated into the analysis for better results. The ideal probability is

$$P(H_i^t | x^t, x^{t-1}, ...)$$

but it is impossible to take into account all of the motion history information. Therefore, we simplify the probability as follows:

$$P(H_i^t | x^t, x^{t-1}, ...) \approx P(H_i^t | H_i^{t-1}, x^t)$$

All of the motion history information is represented by the previous hypothesis $H_i^{(t-1)}$. This result is independent of the evidence x^t. Then,

$$P(H_i^t | H_i^{t-1}, x^t) = P(H_i^t | H_i^{t-1}) P(H_i^t | x^t)$$

where

$$P(H_i|x) = \frac{P(H_i)P(x|H_i)}{P(x)} = \frac{P(H_i)P(x|H_i)}{\sum P(x|H_i)P(H_i)}$$

Therefore, given the starting hypothesis H_i^0 and transfer matrix $P(H_i^t|H_i^{t-1})$, the probability can be induced from the previous results and those of the SVM model.

4.2.2.6 Experiments

An action recognition experiment is carried out to evaluate the method. The experiment is based on a database of five different action types, namely, hitting, kicking, walking, waving, and double waving. The SVM model is trained with about 400 data, and over 280 data are tested. The data are recognized as six action types including the type NOT, which means that the models do not know the action type to which they belong. The recognition results for the training and testing data are shown in Tables 4.2 and 4.3, respectively. The numbers in the tables are the numbers of action data (row) that are classified as different actions (column). It can be seen from the tables that the generalization of the method is acceptable, as the difference between the testing and the training results is small; however, their accuracy needs to be improved.

The better results that are obtained using the Bayesian framework based on similar training and testing data are shown in Tables 4.4 and 4.5, respectively. The recognition rates of different methods under different situations are shown in Figure 4.7. The figure shows that the Bayesian framework-based method largely improves the performance of SVM recognition.

The experiment is carried out on a PC running at 1.7 GHz. Taking advantage of the MHI-based context feature, a speed of about 15 fps can be achieved, which is efficient for real-time surveillance.

Table 4.2 Training data recognition result

Result / Test	NOT	Walk	Wave	DWave	Hit	Kick
Walk	0	46	0	0	0	2
Wave	2	0	75	10	17	5
DWave	0	0	13	81	5	0
Hit	0	0	6	12	77	5
Kick	0	0	2	0	9	40

Table 4.3 Testing data recognition result

Test \ Result	NOT	Walk	Wave	DWave	Hit	Kick
Walk	0	38	1	0	5	0
Wave	0	0	28	6	9	7
DWave	0	0	6	39	9	4
Hit	0	0	6	10	68	8
Kick	0	0	4	0	6	31

Table 4.4 Training data recognition result improved

Test \ Result	NOT	Walk	Wave	DWave	Hit	Kick
Walk	0	48	0	0	0	0
Wave	2	0	87	7	10	3
DWave	0	0	11	85	3	0
Hit	0	0	3	2	91	4
Kick	0	0	0	0	6	45

Table 4.5 Testing data recognition result improved

Test \ Result	NOT	Walk	Wave	DWave	Hit	Kick
Walk	0	41	0	0	3	0
Wave	0	0	36	4	6	4
DWave	0	0	4	46	3	5
Hit	0	0	4	0	80	8
Kick	0	0	8	0	0	33

4.3 Rule-based Behavior Analysis

In some cases, rule-based behavior analysis, although simple, demonstrates robust real-time performance.

It is not easy to track the whole body of a person because of the large range of possible body gestures, which can lead to false tracking. To solve this problem, based on tracking only the upper body of a person (Figure 4.17), we propose a rule-based method that does not vary much with changes in gesture. We take the upper half of the rectangle as the upper body of a target. It may contain some part of the legs or may not contain the whole of the upper body. We can obtain solely the upper body using the clustering method mentioned above. Based on this robust tracking system, we can obtain the speed, height, and width of the target. With the speed of the upper body and the thresholds selected by experiments, running motion can successfully be detected. Also, based on the height and width of the target, we can detect falling motion through shape analysis.

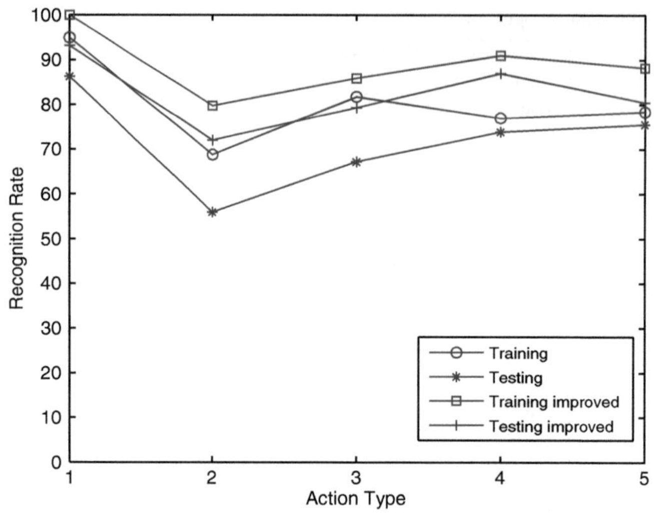

Fig. 4.7 Recognition rate comparison. The x-axis represents different action types: walking (1), waving (2), double waving (3), hitting(4) and kicking (5). The y-axis is the recognition rate (%).

Figure 4.9 shows the detection of abnormal behavior, which includes people falling down, bending down, and running in a household environment, based on the tracking results.

4.4 Application: Household Surveillance Robot

Sometimes, abnormal events occur in a room inside a house or in locations that are outside the field of view of the camera. These events require a different detection approach. How can they be detected? In contrast to video surveillance, audio surveillance does not require that a scene be "watched" directly, and its effectiveness is not influenced by the occlusions that can cause failures in video surveillance systems. Especially in houses or storehouses, many areas may be occluded by moving or static objects.

In this chapter, we present a surveillance system installed on a household robot that detects abnormal events utilizing video and audio information. Moving targets are detected by the robot using a passive acoustic location device. The robot then tracks the targets using a particle filter algorithm. For adaption to different lighting conditions, the target model is updated regularly based on an update mechanism. To ensure robust tracking, the robot detects abnormal human behaviors by tracking the upper body of a person. For audio surveillance, Mel frequency cepstral coefficients (MFCCs) are used to extract features from audio information. Those features are input into a SVM classifier for analysis. The experimental results show that the robot can detect abnormal behavior such as falling down and running. In addition,

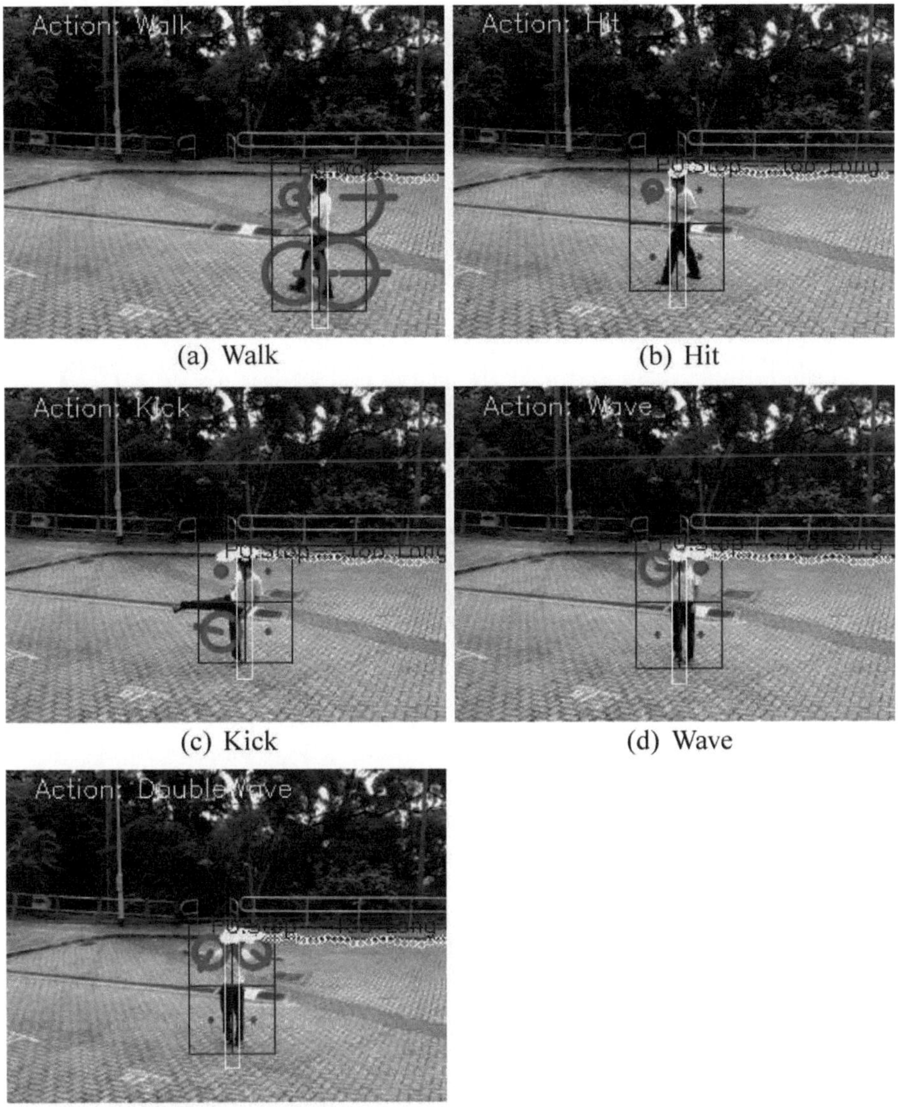

(a) Walk (b) Hit

(c) Kick (d) Wave

(e) Double wave

Fig. 4.8 Action Recognition Result

a 88.17% accuracy rate is achieved in the detection of abnormal audio information such as crying, groaning, and gunfire. To lower the incidence of false alarms by the abnormal sound detection system, the passive acoustic location device directs the robot to the scene where an abnormal event has occurred, and the robot employs its camera to confirm the occurrence of the event. Finally, the robot sends the image that is captured to the mobile phone of its master.

Fig. 4.9 Abnormal behavior detection results.

The testing prototype of the surveillance robot is composed of a pan/tilt camera platform with two cameras and a robot platform (Figure 4.10). One camera is employed to track targets and detect abnormal behaviors. The other is used to detect and recognize faces, which is not discussed in this chapter. Our robot can detect a moving target by sound localization and then track it across a large field of vision using the pan/tilt camera platform. It can detect abnormal behavior in a cluttered environment, such as a person suddenly running or falling down on the floor. By teaching the robot the difference between normal and abnormal sound information, the computational models of action built inside the trained support vector machines can automatically identify whether newly received audio information is normal. If abnormal audio information is detected, then the robot, directed by the passive acoustic location device, can employ its camera to confirm that the event has taken place.

A number of video surveillance systems for detecting and tracking multiple people have been developed, including the W^4 in [162], TI's system in [163], and the system in [164]. Good tracking often depends on correct segmentation. However, among the foregoing systems, occlusion is a significant obstacle. Furthermore, none of them is designed with abnormal behavior detection as its main function. Radhakrishnan et al. [165] presented a systematic framework for the detection of abnormal sounds that might occur in elevators.

Luo [166] built a security robot that can detect dangerous situations and provide a timely alert, which focuses on fire, power, and intruder detection. Nikos [167] presented a decision-theoretic strategy for surveillance as a first step towards automating the planning of the movements of an autonomous surveillance robot.

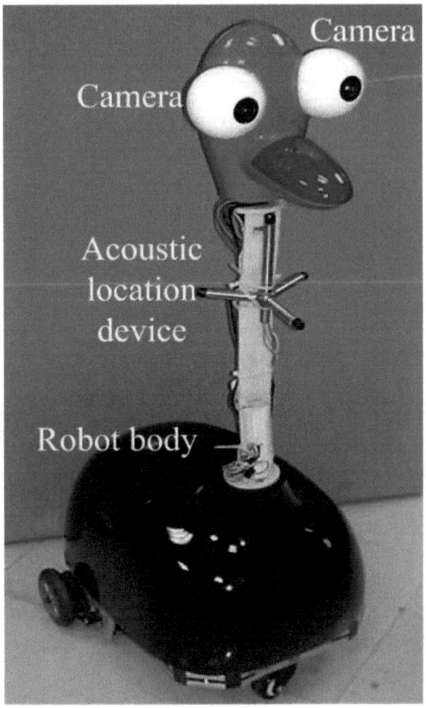

Fig. 4.10 The testing prototype of surveillance robot.

The overview of the system is shown in Figure 4.11. In the initialization stage, two methods are employed to detect moving objects. One is to pan the camera step by step and employ the frame differencing method to detect moving targets during the static stage. The other method uses the passive acoustic location device to direct the camera at the moving object, keeping the camera static and employing the frame differencing method to detect foreground pixels. The foreground pixels are then clustered into labels, and the center of each label is calculated as the target feature, which is used to measure similarity in the particle filter tracking. In the tracking process, the robot camera tracks the moving target using a particle filter tracking algorithm, and updates the tracking target model at appropriate times. To detect abnormal behavior, upper body (which is more rigid) tracking is implemented in a way that focuses on the vertical position and speed of the target. At the same time, with the help of a learning algorithm, the robot can detect abnormal audio information, such as crying or groaning, even in other rooms.

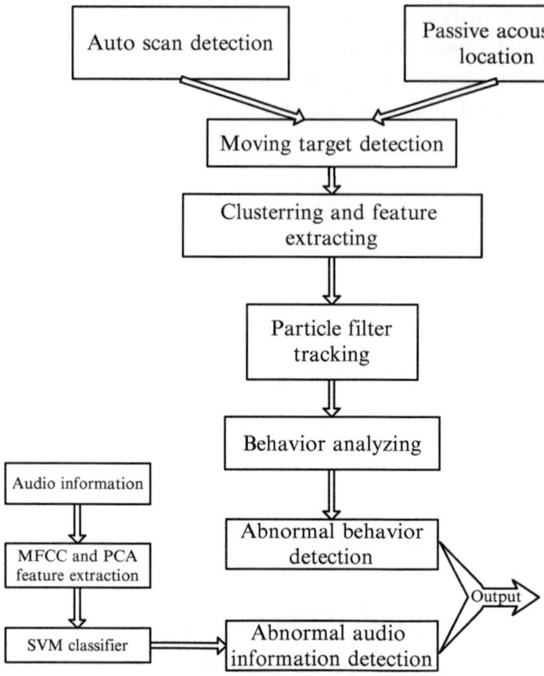

Fig. 4.11 Block diagram of the system.

4.4.1 System Implementation

In many surveillance systems, the background subtraction method is used to find the background model of an image so that moving objects in the foreground can be detected by simply subtracting the background from the frames. However, our surveillance robot cannot use this method because the camera is mobile and we must therefore use a slightly different approach. When a person speaks or makes noise, we can locate the position of the person with a passive acoustic location device and rotate the camera to the correct direction. The frame differencing method is then employed to detect movement. If the passive acoustic location device does not detect any sound, then the surveillance robot turns the camera 30 degrees and employs the differencing method to detect moving targets. If the robot does not detect any moving targets, then the process is repeated until the robot finds a moving target or the passive acoustic device gives it a location signal.

The household robot system is illustrated in Figure 4.12.

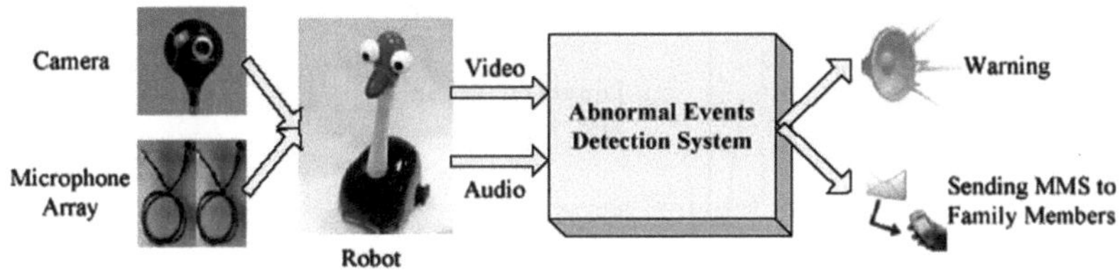

Fig. 4.12 System overview of the house hold robot

4.4.2 Combined Surveillance with Video and Audio

An object producing an acoustic wave is located and identified by the passive acoustic location device. Figure 4.13 shows the device, which comprises four microphones installed in an array. The device uses the time-delay estimation method, which is based on the differences in the arrival time of sound at the various microphones in the sensor array. The position of the sound source is then calculated based on the time delays and the geometric position of the microphones. To obtain this spatial information, three independent time delays are needed; therefore, the four microphones are set at different positions on the plane. Once the direction results have been obtained, the pan/tilt platform moves so that the moving object is included in the camera's field of view.

The precision of the passive acoustic location device depends on the distances between the microphones and the precision of the time delays. In our testing, we find that the passive acoustic location error is about 10 degrees on the x-y plane. The camera angle is about 90 degrees and much greater than the passive acoustic location error (see Figure 4.14). After the passive acoustic location device provides a direction, the robot turns the camera and keeps the target in the center of the camera's field of view. Thus, location errors can be ignored.

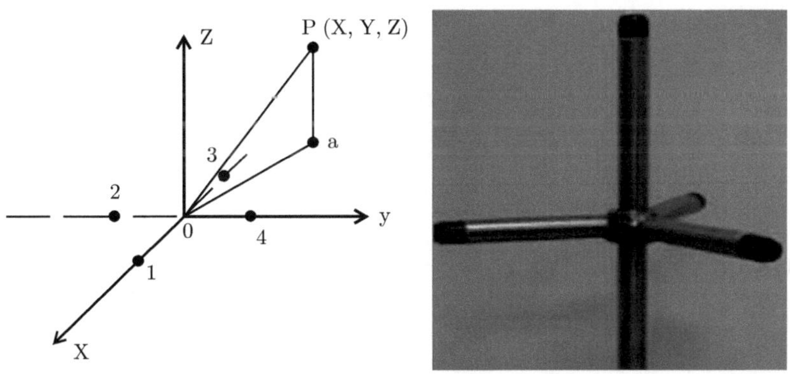

Fig. 4.13 Passive acoustic location device.

In contrast to video surveillance, audio surveillance does not require that the scene be "watched" directly, nor is its effectiveness affected by occlusion, which can

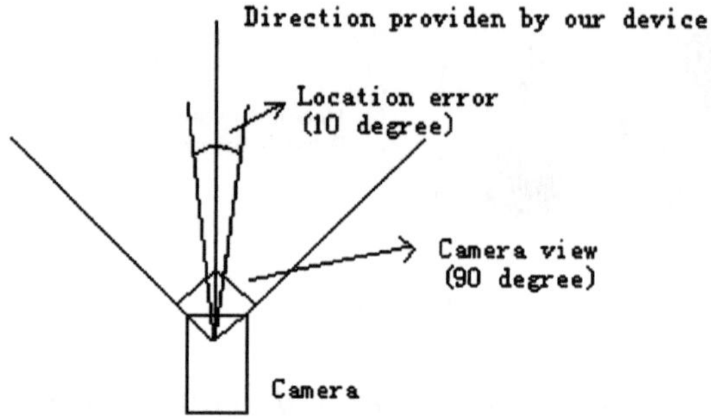

Fig. 4.14 Solving the location error.

result in the reporting of false alarms by video surveillance systems. In many houses or storehouses, some areas may be occluded by moving or static objects. Also, the robot and human operator may not in the same room if a house has several rooms.

We propose a supervised learning-based approach to achieve audio surveillance in a household environment [183]. Figure 4.15 shows the training framework of our approach. First, we collect a sound effect dataset (see Table 4.6), which includes many sound effects collected from household environments. Second, we manually label these sound effects as abnormal (e.g., screaming, gunfire, glass breaking, banging) or normal (e.g., speech, footsteps, shower running, phone ringing) samples. Third, an MFCC feature is extracted from a 1.0 s waveform from each sound effect sample. Finally, we use an SVM to train a classifier. When a new 1.0 s waveform is received, the MFCC feature is extracted from the waveform; then, the classifier is employed to determine whether the sound sample is normal or abnormal.

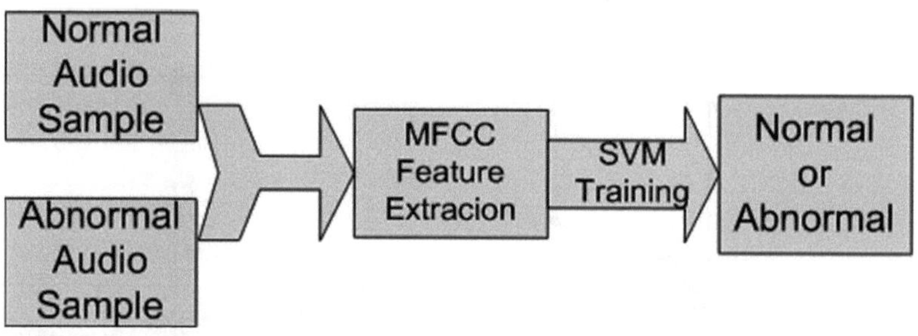

Fig. 4.15 Training framework.

4.4.2.1 MFCC Feature Extraction

To distinguish normal from abnormal sounds, a meaningful acoustic feature must be extracted from the waveforms of sounds.

Table 4.6 The sound effects dataset

Normal sound effects	Abnormal sound effects
normal speech	gun shot
boiling water	glass breaking
dish-washer	screaming
door closing	banging
door opening	explosions
door creaking	crying
door locking	kicking a door
fan	groaning
hair dryer	
phone ringing	
pouring liquid	
shower	
...	...

Many audio feature extraction methods have been proposed for different audio classification applications. The spectral centroid, zero-crossing rate, percentage of low-energy frames, and spectral flux [172] methods have been used for speech and music discrimination tasks. The spectral centroid represents the balancing point of the spectral power distribution; the zero-crossing rate measures the dominant frequency of a signal; the percentage of low-energy frames describes the skewness of the energy distribution; and the spectral flux measures the rate of change of the sound. Timbral, rhythmic, and pitch [173] features, which describe the timbral, rhythmic, and pitch characteristics of music, respectively, have been proposed for automatic genre classification.

In our approach, the MFCC feature is employed to represent audio signals. Its use is motivated by perceptual and computational considerations. As it captures some of the crucial properties of human hearing, it is ideal for general audio discrimination. The MFCC feature has been successfully applied to speech recognition [174], music modeling [175], and audio information retrieval [176]. More recently, it has been used in audio surveillance [177].

The steps to extract the MFCC feature from the waveform are as follows.

Step 1: Normalize the waveform to the range $[-1.0, 1.0]$ and apply a hamming window to the waveform.

Step 2: Divide the waveform into N frames, i.e., $\frac{1000}{N}$ ms for each frame.

Step 3: Take the fast Fourier transform (FFT) of each frame to obtain its frequency information.

Step 4: Convert the FFT data into filter bank outputs. Since the lower frequencies are perceptually more important than the higher frequencies, the 13 filters allocated below 1000 HZ are linearly spaced (133.33 Hz between center frequencies) and the 27 filters allocated above 1000 Hz are spaced logorithmically (separated by a factor of 1.0711703 in frequency). Figure 4.16 shows the frequency response of the triangular filters.

Step 5: As the perceived loudness of a signal has been found to be approximately logarithmic, we take the log of the filter bank outputs.

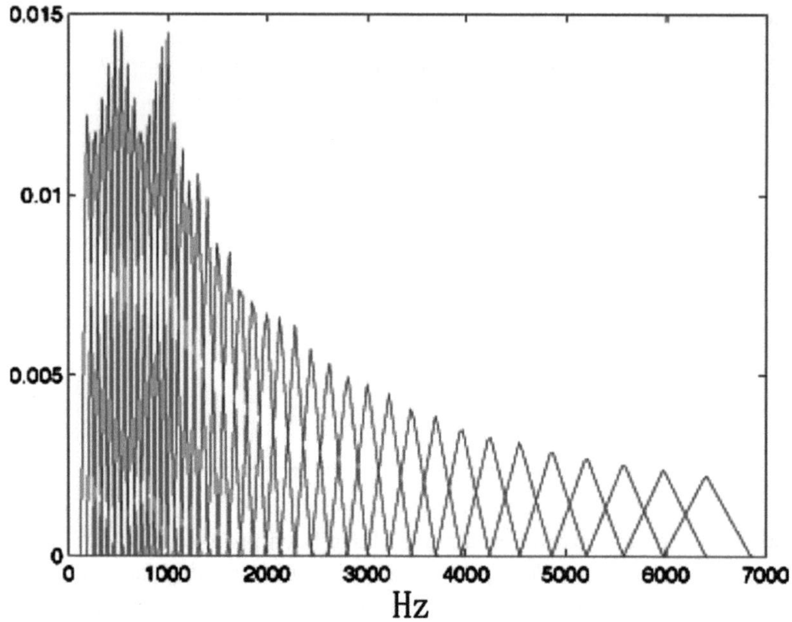

Fig. 4.16 Frequency response of the triangular filters.

Step 6: Take the cosine transform to reduce dimensionality. As the filter bank outputs that are calculated for each frame are highly correlated, we take the cosine transform that approximates PCA to decorrelate the outputs and reduce dimensionality. Thirteen (or so) cepstral features are obtained for each frame using this transform. If we divide the waveform into 10 frames, then the total dimensionality of the MFCC feature for a 1.0 s waveform is 130.

4.4.2.2 Support Vector Machine

After extracting the MFCC features from the waveforms, we employ an SVM-trained classifier to determine whether this sound is normal or not [178][179][180].

Our goal is to separate sounds into two classes, normal and abnormal, according to a group of features. Many types of neural networks are used for binary classification problems, including SVMs, RBFNs, the nearest neighbor algorithm, Fisher's linear discriminant, and so on. We choose an SVM as the audio classifier because it has stronger theory interpretation and better generalization performance than neural networks. Compared with other neural network classifiers, SVM ones have three distinct characteristics. First, they estimate a classification using a set of linear functions that are defined in a high-dimensional feature space. Second, they carry out classification estimation by risk minimization, where risk is measured using Vapnik's ε-insensitive loss function. Third, they are based on the structural risk minimization (SRM) principle, which minimizes the risk function, which consists of the empirical error and a regularized term.

4.4.3 Experimental Results

To evaluate our approach, we collect 169 sound effect samples from the Internet [184], including 128 normal and 41 abnormal samples, most of which are collected in household environments.

For each sample, the first 1.0 s waveform is used for training and testing. The rigorous jack-knifing cross-validation procedure, which reduces the risk of overstating the results, is used to estimate classification performance. For a dataset with M samples, we chose $M-1$ samples to train the classifier, and then test performance using the remaining sample. This procedure is then repeated M times. The final estimation result is obtained by averaging the M accuracy rates. To train a classifier using SVM, we apply a polynomial kernel, where the kernel parameter d = 2, and the adjusting parameter C in the loss function is set to 1.

Table 4.7 shows the accuracy rates using the MFCC feature trained with different frame sizes, while Table 4.8 shows those using MFCC and PCA features trained with different frame sizes. PCA of the dataset is employed to reduce the number of dimensions. For example, the number is reduced from 637 to 300 when the frame size is 20 ms and from 247 to 114 when the frame size is 50 ms. Comparing the results in Tables 4.7 and 4.8, we find that it is unnecessary to use PCA. The best accuracy rate obtained is 88.17% with a 100 ms frame size. The number of dimensions of this frame size is 117 and thus dimension reduction is unnecessary.

Nowadays, abnormal behavior detection is a popular research topic, and many studies have presented methods to detect abnormal behaviors. Our surveillance robot is for use in household environments, so it needs to detect abnormal behaviors such as falling down or running.

Table 4.7 Accuracy rates (%) using the MFCC feature trained with 20, 50, 100, and 500 ms frame sizes.

Frame size	20 ms	50 ms	100 ms	500 ms
Accuracy	86.98	86.39	88.17	79.88

Table 4.8 Accuracy rates (%) using the MFCC and PCA features trained with 20, 50, 100, and 500ms frame sizes.

Frame size	20 ms	50 ms	100 ms	500ms
Accuracy	84.62	78.70	82.84	76.33

It is not easy to track the whole body of a person because of the large range of possible body gestures, which can lead to false tracking. To solve this problem, we propose a method that only tracks the upper body of a person (Figure 4.17), which does not vary much with gesture changes. We take the upper half rectangle as the upper body of a target. It may contain some part of legs or lost some part of upper body. We can obtain pure upper body by using the clustering method mentioned before. Based on this robust tracking system, we can obtain the speed of the target, the height and width of the target. Through the speed of the upper body and the thresholds selected by experiments, running movement can be successfully detected. Also, based on the height and width of the target, we can detect falling down movement through space analysis.

Figures 4.18 (a)-(d) show the robot moving in the proper direction to detect the target upon receiving a direction signal from the passive acoustic location device. Figures 4.18 (a) and (b) are blurry because the camera is moving very quickly, whereas Figures 4.18 (e) and 4.18 (f) are clear as the robot tracks the target and keeps it in the center of the camera's field of view.

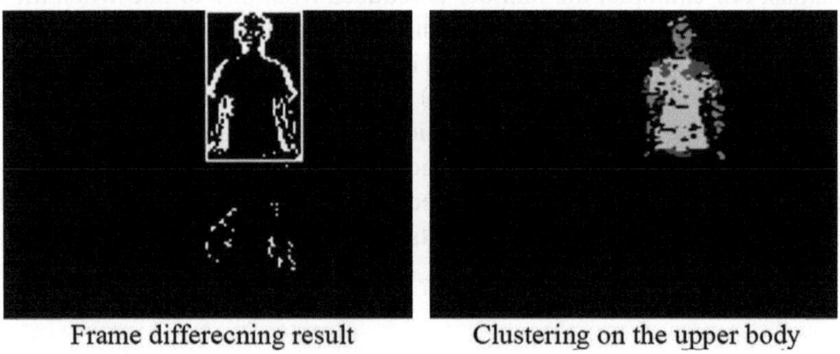

Fig. 4.17 Clustering results on the upper body.

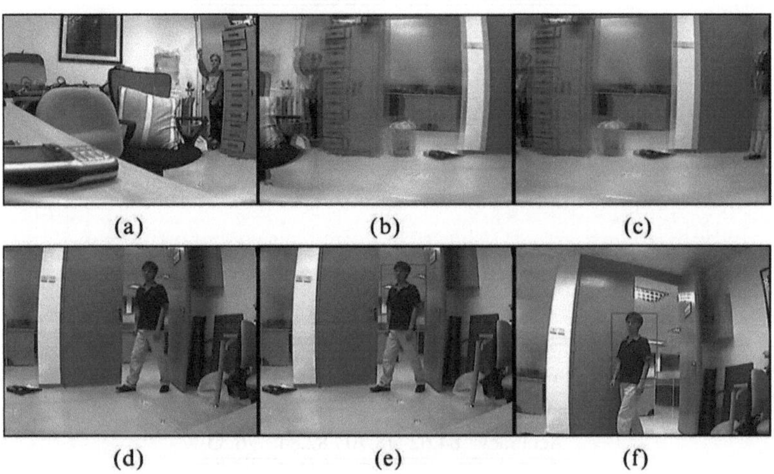

Fig. 4.18 Initialization and tracking results.

We test our abnormal sound detection system in a real-world work environment. The sounds include normal speech, a door opening, the tapping of keys on a keyboard, and so forth, which are all normal sounds. The system gives eight false alarms in one hour. A sound box is used to play abnormal sounds, such as gunfire and crying, among others. The accuracy rate is 83% among the 100 abnormal sound samples, which is lower than that obtained in previous noise experiments conducted in real-world environments.

To lower the incidence of false alarms by the abnormal sound detection system, the passive acoustic location device directs the robot to the scene where an abnormal event has occurred and the robot employs its camera to confirm the occurrence of the event. In Figure 4.19, we can see the process of abnormal behavior detection utilizing video and audio information when two persons are fighting in another room. First, the abnormal sound detection system detects abnormal sounds (fighting and shouting). At the same time, the passive acoustic location device obtains the direction. The robot then turns towards the direction whence the abnormal sound

(a) (b)

The images are captured by us using a digital video

(c) (d)

The images are captured by the robot

Fig. 4.19 The process of abnormal behavior detection utilizing video and audio information: (a) The initial state of the robot; (b) The robot turns to the direction where abnormality occurs; (c) The initial image captured by the robot; (d) The image captured by the robot after turning to direction of abnormality.

Fig. 4.20 The robot sends the image to the mobile phone of its master.

originates and captures images to check whether abnormal behavior is occurring. To detect fighting, we employ the optical flow method, which we have discussed in the previous chapter [182].

Some abnormal cases cannot correctly be detected by our system. One example is the case where a person has already fallen before the robot turns to the correct direction. To solve this problem, the robot sends the image captured to the mobile phone of its master (Figure 4.20).

4.5 Conclusion

In this chapter, we present the behavior analysis of the actions of individuals and groups. These cases have one thing in common - the environment is not crowded, so clear human blobs are available for detailed behavior analysis. A learning-based approach is demonstrated to be effective in the behavior analysis of individuals. We develop both a contour- and a motion-based method of behavior analysis. A rule-based method, because of its simplicity, also works for the classification of unsophisticated behaviors.

In this chapter, we describe a household surveillance robot that can detect abnormal events by combining video and audio surveillance techniques. Our robot first detects a moving target by sound localization, and then tracks it across a large field of vision using a pan/tilt camera platform. The robot can detect abnormal behaviors in a cluttered environment, such as a person suddenly running or falling down on the floor. It can also detect abnormal audio information and employs a camera to confirm events.

This research offers three main contributions: a) an innovative strategy for the detection of abnormal events that utilizes video and audio information; b) an initialization process that employs a passive acoustic location device to help a robot detect moving targets; and c) an update mechanism to regularly update the target model.

Chapter 5
Facial Analysis of Individuals[1]

In the previous chapter, we presented behavior analysis, which is based on dynamic information, including background extraction based on a series of frames, blob detection based on two consecutive frames, and so forth. However, for intelligent surveillance, we can also explore static information, i.e., facial images. Facial detection is straightforward as it does not depend on whether or not background information is extractable. Hence, it is useful in highly cluttered environments, with the only requirement being sufficient image resolution of the image.

Much research has been conducted into the identification of humans based on facial images in real time. In addition to facial recognition, the binary classification of facial images has recently attracted much attention. A successful face classification algorithm can significantly boost the performance of other applications, such as facial recognition, and can also be used as a smart human-computer interface.

Despite being widely investigated in the field of psychology [80] and [81], race classification has received less attention in computer vision studies compared to gender or age classification. Gollomb et al. [82] trained a two-layer neural network, SEXNET, to determine gender from facial images, and reported an average error rate of 8.1% based on 90 images (45 females and 45 males). Brunelli and Poggio [83] developed a method based on HyperBF networks, and the results for a dataset of 168 images showed an average error rate of 21%. Sun et al. [85] used genetic feature subset selection for gender classification, PCA for feature generation, and a genetic algorithm for feature selection. They achieved an accuracy rate of 86.3% based on a database of 300 images. However, it is difficult to evaluate the relative performance of these systems, as the image databases used in each case are different. Moghaddam et al. [84] investigated race classification using SVMs for a database of 1755 facial images from FERET , and reported an accuracy of 96.6%. Shakhnarovich et al. [91] introduced a unified learning framework for face classification, which uses the so-called rectangle features for classification. They tested the

[1] Portions reprinted, with permission, from Yongsheng Ou, Xinyu Wu, Huihuan Qian, and Yangsheng Xu, A real time race classification system, *2005 IEEE International Conference on Information Acquisition*. ©[2005] IEEE.

H. Qian et al., *Intelligent Surveillance Systems*, Intelligent Systems, Control and Automation: Science and Engineering 51, DOI 10.1007/978-94-007-1137-2_5,

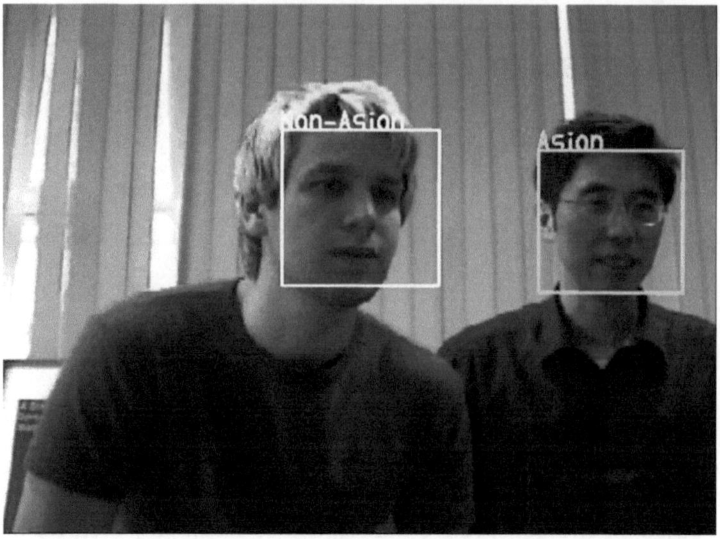

Fig. 5.1 Two detected faces and the associated classification (Asian/non-Asian) results.

system with unnormalized faces taken directly from the face detection system [92], and achieved a classification rate of 78%.

Our system (see Figure 5.1) shares some properties with that of [91], in that we do not apply a face alignment algorithm, and test our face classification system in an unconstrained environment. We have rejected alignment because it diminishes the robustness and efficiency of the system, as it requires the automatic extraction of facial features, a process that is not entirely robust in the presence of great variation in poses, lighting, and quality.

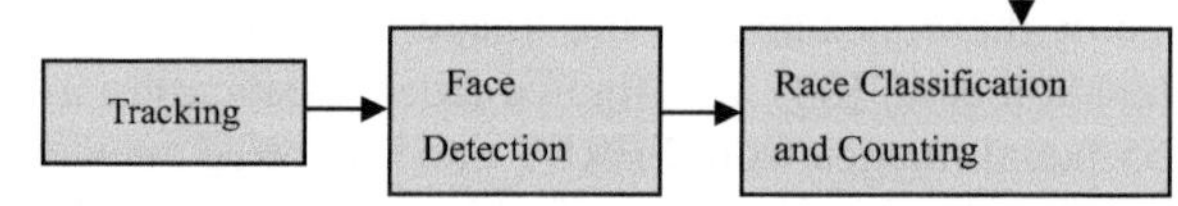

Fig. 5.2 Block diagram of the system

The overview of our system is shown in Figure 5.2. In the training stage, PCA and ICA are employed to generate and select features from the images, and the classifier is trained by an SVM. After obtaining the classifier, we track multiple people using distance and color tracking algorithms [86], and then detect faces based on the tracking results. In other words, we do not detect faces in the whole image but only in the tracked blobs, which saves time and avoids the detection of the wrong face. Our system makes two major contributions. (1) We use ICA to map the features produced by PCA to generate new features, which has proven to be more suitable for the classifier. (2) We use data fusion techniques to combine different classifiers and boost the classification rate to a new level.

The remainder of this chapter is organized as follows. Section 5.1 describes the generation and selection of features from facial images by combining PCA and ICA algorithms. In section 5.2, we apply a combination of SVM classifiers to boost the classification rate, and in section 5.3, we present the experimental results. We summarize the chapter in section 5.4.

5.1 Feature Extraction

Feature generation and selection are very important in face classification, as poorly selected features, such as gender, identity, glasses, and so on, can obviously reduce the level of performance. In addition, although certain features may contain enough information about the output class, they may not predict the output correctly because the dimensions of the feature space may be so large that numerous instances are required to determine the result. In our system, we use a supervised feature selection method that combines PCA and ICA algorithms.

5.1.1 Supervised PCA for Feature Generation

There are many feature generation methods for information packing, including singular value decomposition (SVD), discrete Fourier transform (DFT), and discrete cosine transform (DCT). The Karhunen-Loeve (KL) transform, also known as PCA, can generate mutually uncorrelated features while packing most of the information into several eigenvectors. It is widely used in pattern recognition and in a number of signal and image processing applications. PCA-based methods have also been proven to be more powerful in interpreting information from faces than other dimension reduction techniques.

In the eigenface method for facial recognition, which was introduced by Turk and Pentland [87], a set of eigenvectors is computed from training faces, some of which are then selected for classification according to their eigenvalue. It was reported that different eigenvectors encode different information, with eigenvectors 1-4 encoding light variations and eigenvectors 10 and 20 encoding information about glasses. However, no detailed investigation of the relation between eigenvectors and the features of a human face has been conducted. In another approach, which is based on information theory, every feature is considered to encode different information on a certain scale, such as gender, race, or identity.

In most previous studies, PCA is implemented in an unsupervised way. That is, either all of the features generated by PCA are used for the classification task or some features are selected according to the corresponding eigenvalues. In this chapter, we use supervised PCA as a preprocessing step of ICA. We select the eigenvectors according to their binary classification capability, instead of the corresponding eigenvalues. Eigenvectors with large eigenvalues may carry common features, but not the information that allows us to distinguish the two classes. The method is described in brief as follows. Suppose that we have two sets of training samples, A and B. The number of training samples in each set is N. Φ_i represents each eigenvector produced by PCA. Each of the training samples, including positive and negative samples, can be projected onto an axis extended by the corresponding eigenvector. By analyzing the distribution of the projected $2N$ points, we can roughly select the eigenvectors that have more face classification information. The following is a detailed description of the process.

1. For a certain eigenvector Φ_i, compute its mapping result according to the two sets of training samples. The result can be described as $\lambda_{i,j}$, $(1 \leq i \leq M, 1 \leq j \leq 2N)$.
2. Train a classifier f_i using a simple method such as a perception or neural network, which can separate $\lambda_{i,j}$ into two groups, Asian and non-Asian, with a minimum error $E(f_i)$.
3. If $E(f_i) > \theta$, then we delete this eigenvector from the original set of eigenvectors.

M is the number of eigenvectors, $2N$ is the total number of training samples, and θ is the predefined threshold. The remaining eigenvectors are selected. The eigenvectors can also be represented as facial images, which are called eigenfaces.

The evaluation of the eigenvectors is shown in Figure 5.3. The performance of Eigenvector i is better than that of either Eigenvector 1 or 2.

It is possible to select too few or even no good eigenvectors in a single PCA analysis process. We propose the following approach to solve this problem. We assume

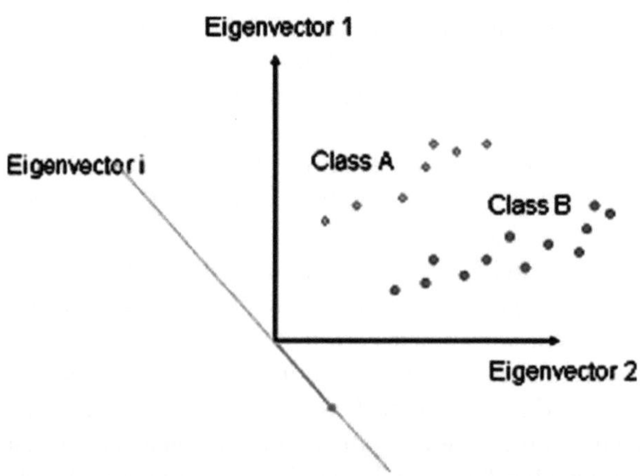

Fig. 5.3 Relative distinguishing ability of the different eigenvectors.

that the number of training samples, $2N$, is sufficiently large. We then randomly select training samples from the two sets. The number of selected training samples in each set is less than $N/2$. We perform supervised PCA with the selected samples. By repeating the previous process, we can collect a number of good features. This approach is inspired by the bootstrap method, the main idea of which is to emphasize some good features by reassembling data, which allows the features to stand out more easily.

5.1.2 ICA-based Feature Extraction

As we can see from the foregoing descriptions, the information contained in eigenvectors is not absolutely independent. It can confuse the classifier and lessen the efficiency of the whole system. In traditional algorithms, such as genetic or branch-and-bound algorithms, features are directly selected using different feature selection methods. Our method maps the eigenvectors that are produced by PCA onto another space using ICA, where the mapped values are statistically independent. In this way, we can provide classifiers with a clearer description of features.

ICA has recently emerged as a powerful solution to the problem of blind signal separation ([88] and [89]), and its use for facial recognition has been proposed by Barlett and Sejnowski [90]. The basic idea of ICA is to take a set of observations and find a group of independent components that explain the data. PCA considers only second-order moments and uncorrelates the data, whereas ICA accounts for higher-order statistics and thus provides a more powerful expression of the data than does PCA.

Figure 5.4 shows a comparison between PCA and ICA in feature extraction for the face classification task. In the figure, the x-axis represents the index of a certain

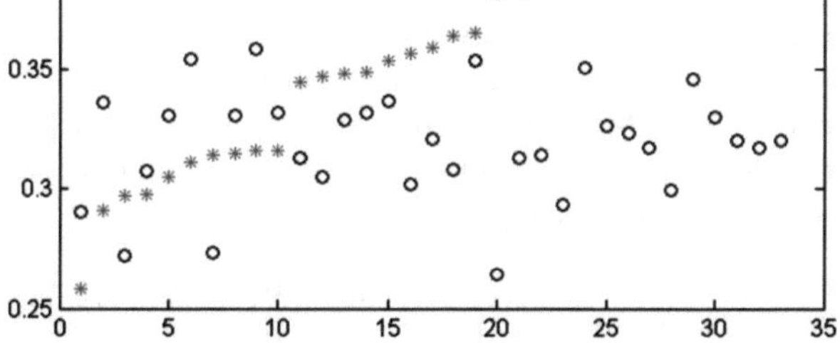

Fig. 5.4 Comparison between PCA and ICA.

feature, and the y-axis represents the corresponding discriminability in performance evaluation. We perform ICA for the 33 good features produced by PCA, which yields 33 ICA features that are statistically independent. The distinguishing capability of both PCA and ICA features is then computed, as shown in Figure 5.4. The figure reveals that although several ICA features with a lower distinguishing rate are generated, ICA features with a higher distinguishing rate are also produced. Several of the latter can thus be selected as the features for use in the classification. It is clear that the method allows us to collect better features.

5.2 Fusion of SVM Classifiers

Our goal is to separate facial images into two classes, Asian and non-Asian, according to a group of features. Many types of learning algorithms can be used for such a binary classification problem, including SVMs, RBFNs, the nearest neighbor algorithm, and Fisher's linear discriminant. The widely used SVM method has been found to have superior classification ability in some studies [84]. Hence, we choose SVMs as our face classifiers.

The PCA and ICA feature extraction and selection process gives us several sets of good feature groups. Using different combinations of these groups, we can obtain a large number, K, of SVM classifiers by learning from a set of training data. Each of the SVM classifiers has a different classification rate and different features. As the number of inputs for a SVM model cannot be very large, any individual SVM classifier has a limited amount of discriminant information. In this section, we address how they can be combined to produce a more powerful face classifier with more features.

Viola and Jones in [92] proposed the AdaBoost algorithm, which initially seems promising for our problem. However, when we add classifiers or features one by one, the classification rate does not continuously increase, but varies up and down. It is thus not always true that using more classifiers (or features) produces better classification rates. The classification rate is almost flat when we combine the good features more than 200 times. Thus, the AdaBoost algorithm may be suitable for rectangular features, but not for our PCA-based ones.

For our problem, we propose a new algorithm, which we call the "Three to One" or "321" algorithm. It is explained as follows.

1. Work out the classification results R_i for all of the available SVM classifiers in all of the labeled sample facial images in the set $\{1,-1\}$.
2. Combine every three SVM classifiers to form a new classifier by summing their results, and decide the final result for the new 321 classifier according to the sign of the result for each labeled sample image.

$$R_n = \begin{cases} 1 & \text{if } (R_1+R_2+R_3) \geq 1 \\ -1 & \text{if } (R_1+R_2+R_3) \leq -1 \end{cases} \tag{5.1}$$

where R_1, R_2, and R_3 are the results for a sample facial image of three different classifiers, and R_n is the result for that facial image for a new classifier, which was produced by fusing the three initial classifiers.

3. Calculate the classification rate ρ of each new 321 classifier according to

$$\rho = \frac{n_c}{n_t}, \tag{5.2}$$

where n_c is the number of correctly classified samples, and n_t is the total number of the samples.

4. Remove the new 321 classifier if its classification rate ρ is less than a predefined threshold ρ^*.
5. Repeat Steps 2-4 with the new 321 classifiers to produce the next level of 321 classifiers until a usable classifier is created or the number of iterations passes a threshold N_r.

This is an extremely time-consuming selection process as it must cover all combinations of the three classifiers in the current level. We need therefore to define a suitable threshold ρ^* as a compromise between reducing computation time and producing good fusion classifiers that perform better.

We propose the use of "321" classifiers rather than "521," "721," or even more, as it would take too much time for a computer to complete the work for the latter types. For example, if 200 classifiers were available, then it would take a computer 10 minutes to work out the next level of 321 classifiers, whereas it would spend about 13 days completing the 521 classifiers.

To avoid overfitting, the following rules should be followed.

1. The sample facial images should be different from the samples for training SVM classifiers.
2. The sample facial images should be representative and well proportioned.
3. The number of sample facial images should be sufficiently large.
4. The number of repeats, N_r, should be decided by the number of samples, and cannot be too large.

A typical example is shown in Figure 5.5.

In the structure of the example, the 27 SVM classifiers are independently trained, but some of them may be the same as very strong features may appear several times.

To summarize, we suggest the use of the 321 algorithm in place of other neural networks as another form of SVM learning, for the following reasons.

1. The 321 algorithm in Equation (5.1) is much faster than most neural network structures in the training process. If there are K classifiers available, then C_K^3 time is needed for training.
2. For 3-input-1-output training, the learning precision of the 321 algorithm is similar to that of SVMs.
3. It is likely that overfitting will result if we choose neural networks with complex structures.

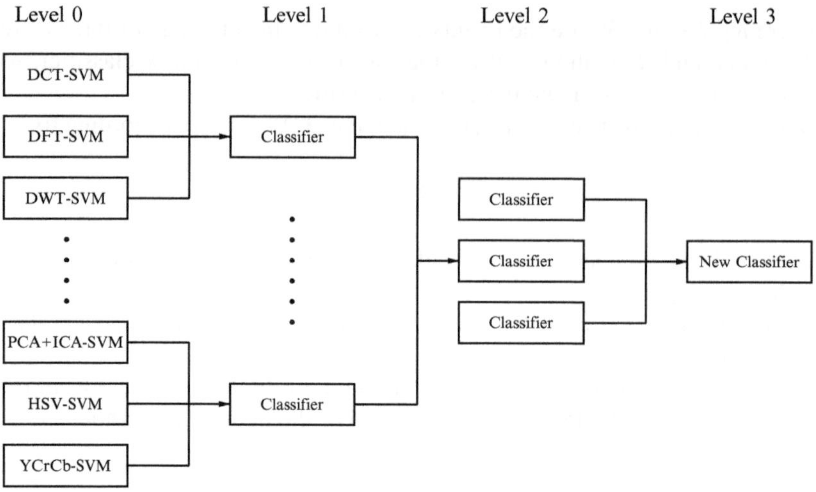

Fig. 5.5 A "321" classifier structure.

5.3 System and Experiments

The facial classification algorithm has been embedded into an intelligent painting system, as shown in Figure 5.6. The kernel controller is a PC104, with a Celeron CPU and 256 MB of RAM. The operating system is Windows 2000. The peripheral circuits include two sets of 7.2 V NiMH batteries (one for the PC104 and the other for the loudspeaker), a power regulation circuit, a digital camera to capture images, and a loudspeaker. The objective application of this system is to greet guests to a museum and introduce them to exhibits. It selects the language to be used, English or Chinese, based on facial information analysis.

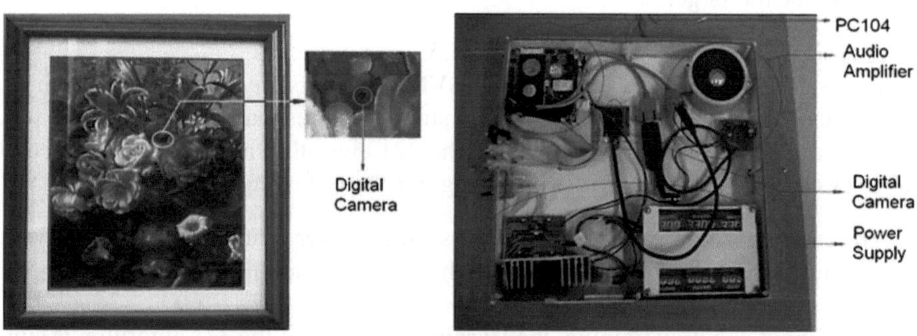

Fig. 5.6 Intelligent Painting for Face Classification

5.3.1 Implementation

To estimate the performance of the proposed system, we collect images of human faces from a large Internet database that contains more than 9000 facial images. The faces are detected automatically using the frontal face detection system [92]. We manually label the results. As the faces are extracted from unconstrained pictures or video, they can be of poor quality. Such faces are rarely completely frontal or may be weakly aligned. In addition, lighting and image quality vary widely. Figure 5.7 shows some sample facial images from the output of the face detection system.

Fig. 5.7 Sample images extracted from the output of face detection system.

All of the images are resized into a standard image size of 24×24 pixels. We divide all of the images in the database into two folders, each of which contains both Asian and non-Asian facial images. The first folder of the face database is used for feature extraction, SVM classifier training, evaluation, and testing of the classifiers in the first step, and the second one is used for training and testing the fusion classifiers in the second step.

Based on the first data folder, we test different feature extraction methods, including PCA and ICA. Table 5.1 presents the results. The initial ones show that the supervised ICA method outperforms the others, whereas the final ones show that the PCA + ICA method outperforms the others.

Feature extraction	non-Asian	Asian	Total error
PCA	37.1%	44%	40.9%
Supervised PCA	38.1%	37.4%	37.6%
PCA+ICA	35.8%	38.4%	36.1%

Table 5.1 Sample human control data.

In the second step, we combine different SVM classifiers using the 321 algorithm. For comparison, we test several of the most successful combined classifiers with the best original SVM classifiers. The results, which are presented in Table 5.2, show that combining the different classifiers improves the whole classification rate.

Classifier	non-Asian	Asian	Total error
SVM	22%	24.6%	23.7%
level 1	20.9%	21.5%	20.7%
level 2	18.2%	19.2%	18.7%
level 3	17.3%	16.8%	17%

Table 5.2 "321" testing result.

5.3.2 Experiment Result

In testing the classification process, the FERET database [93] is used as a benchmark database. The tested facial images are extracted automatically using the frontal face detection system [92]. In all, there are 750 Asian and non-Asian faces. All of the facial images are resized into 24×24 pixel images. None has previously been used to train or test the system. Table 5.3 shows the results.

Classifier	non-Asian	Asian	Total error
Fusion Classifier	16.1%	22.3%	17.5%

Table 5.3 FERET database testing result

5.4 Conclusion

In this chapter, a new real-time face classification system based on static facial images is presented, which could be an important branch of intelligent surveillance. We put selected PCA and ICA features into SVM classifiers, and by fusing the classifiers achieve a good classification rate. Our system can also be used for other types of binary classification of facial images, such as gender or age classification, with only little modification.

Chapter 6
Behavior Analysis of Human Groups[1]

6.1 Introduction

In the surveillance of public areas, human groups should be the major targets monitored because of their highly sociable character. In groups, the behavior of individuals usually depends on that of others. Hence, group behavior can explain a large number of phenomena, including overcrowding, chaos, stampedes (as a result of a terrorist attack such as a bomb explosion), fights, and riots.

Some researchers [198, 199] have attempted to deal with the this category of tasks. They have conducted research into multi-object/-target tracking from different angles but neglected the relationship within the overlapped people group.

Taking inspiration from the concept of the agent in artificial intelligence, we develop a multi-agent method to track and analyze groups of people and detect events. Although the notion of an agent is not new in the area of computer vision, researchers understand and use the term differently.

For instance, in [201], an agent refers both to the object in the field of view, i.e., a person, or to the instrument used to track it, i.e., the camera, while in [202], an agent denotes both the camera and its combination with the backstage processing program capable of camera control and image processing. In [200], agents are persons or vehicles in a parking lot, and the authors focus on the person-vehicle relationship. Finally, in [203], an agent is a virtual entity that is used for human modeling and simulation.

Much of the research into surveillance systems uses a multi-agent approach. Adopting the CHHM approach, [204] successfully solved the problem of interaction between two persons. [205] also focused on two-object interaction. [206] used the pair primitives method to investigate surveillance based on the interrelationship between two primitives (the word "primitive" means roughly the same as agent as

[1] Portions reprinted, with permission, from Yufeng Chen, Zhi Zhong, Ka Keung Lee, and Yangsheng Xu, Multi-agent Based Surveillance, *2006 IEEE/RSJ International Conference on Intelligent Robots and Systems*. ©[2006] IEEE.

H. Qian et al., *Intelligent Surveillance Systems*, Intelligent Systems, Control and Automation: Science and Engineering 51, DOI 10.1007/978-94-007-1137-2_6,

discussed in this chapter). All of the agent or multi-agent approaches mentioned above can only be employed in the surveillance of a single person or a pair of persons, respectively, and thus are inadequate for modeling the surveillance of people in public places.

In this section, an agent represents the state of an individual of a group in a field of view and the agent's interaction with other individuals. By computing the interrelationships and interactions of more than two agents, surveillance for those events that take place which involve groups of people can be realized. This function constitutes the basis for the realization of intelligent surveillance.

The term "agent" is borrowed from artificial intelligence and game theory [207], and represents an artificial object that has independent status and actions. In our surveillance system, an agent records the basic parameters and infers the status of an object for further group analysis.

Information about agent G could be viewed as a status in terms of time t, where status Θ_t is a vector of basic parameters, including location x_t, y_t, height h_t, and even the object's template T_t.

The inferential status of an agent, which is also a part of the vector Θ_t, is updated at each step. All of the single agent-wide information can be inferred from the basic parameters, varying from low-level information such as speed V_t and direction d_t to high-level information such as action type A_t.

The basic structure of agent-based behavior analysis is shown in Figure 6.1. The agent is driven by the image data at time t, and all of the information in the previous steps can be used to help track the status of the next step. Finally, different levels of certain kinds of events can be detected by statistical analysis of the status of all agents.

6.2 Agent Tracking and Status Analysis

As seen in the segmentation and tracking chapter, the body of a person is often occluded by others, and thus detecting the whole body is impossible under certain situations. The head becomes the best feature for detection and tracking, since it can usually be seen, even in a crowd.

A hybrid tracking method is used as shown in Figure 6.2. We adopt the contour-based method [208], which is very efficient in the detection of similar contour features after background subtraction. In addition, the template-matching method is combined with the Kalman filter for tracking as the contour feature cannot always be detected in the crowd. Each time the head contour appears, as shown in Figure 6.3(a) 6.3(b), the head region is detected and the head image template updated. In case the contour cannot be detected in the next step, as in Figure 6.3(c) 6.3(d), the image template from the last step is used to search for the most similar region in the neighborhood.

Once the head's position is tracked, much basic information about direction and speed can be derived, and image analysis can be done based on the information to obtain the person's height, head template, and action type.

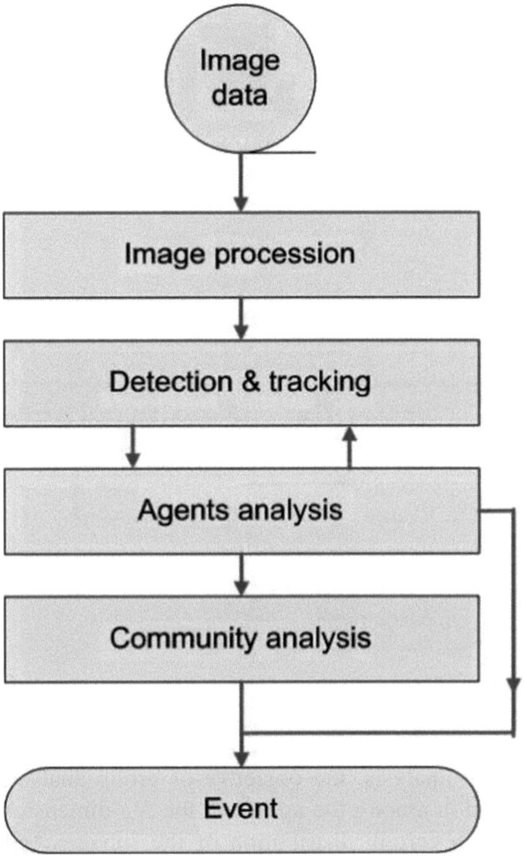

Fig. 6.1 Surveillance structure.

Moreover, as the parameter varies with the time t, some kinds of agent movement such as walking, running, jumping, and stopping can be recognized by the vector Θ_t or its transformed version.

We use an SVM [209], which has been trained with some selected samples, on the Harr wavelet coefficients of the status vector to recognize the actions. More information about the process can be found in the related literature; hence, we do not go into detail about it here. We further discuss multi-agent analysis in the following sections.

6.3 Group Analysis

For each moment t, an agent set $S_t = \{\Theta_t^1, \Theta_t^2, \Theta_t^3 ... \Theta_t^n ... \Theta_t^{N_{st}}\}, 1 \leq n \leq Ns_t$ can be retrieved, where Ns_t is the agent amount, and Θ_t^n is a $Ng \times 1$ vector $\Theta_t^n = (x_t^n, y_t^n, h_t^n, V_t^n, d_t^n ...)^T$.

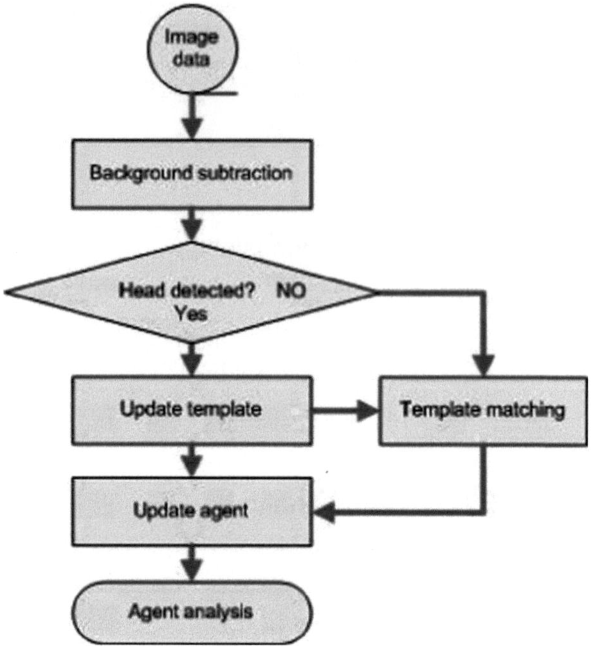

Fig. 6.2 Tracking scheme

In contrast to agent analysis, the objective of group analysis is to obtain the characteristic distribution among the agents in the Ng-dimensional space. We define a group event as a certain distribution of the subagent set in the substatus space.

Regarding the data space, the dimension Ng may be very large as many parameters are used to describe the agent, including the image template, so that dimension reduction must be performed to simplify the problem. In addition, some of the parameters in status Θ_t^n are independent of each other and have a different meaning or level of importance. The usefulness of general methods such as PCA and ICA is thus limited in this application, because such approaches need a large amount of data for training and treat each dimension the same way.

For the example shown in Figure 6.4(a), the points of an agent are displayed in the space extended by the angle and location y, which are selected specifically, as shown in Figure 6.4(b). Based on the angle dimension, the agent set should be divided into two subsets $\{A,D\},\{B,C\}$, and based on the location dimension, the subset should be $\{A,B\},\{C,D\}$. Both schemes have physical meaning, and neither is obviously better than the other. That depends on the application requirements.

It is better to start the analysis from a certain independent subspace such as a location, velocity, or action type. For more complex events, the space can be extended by these perpendicular subspaces.

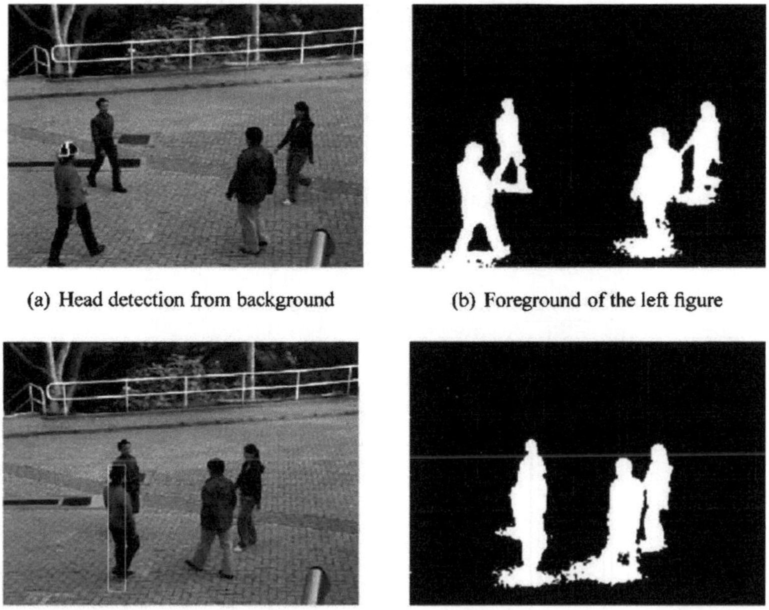

(a) Head detection from background (b) Foreground of the left figure

(c) Head tracked by template (d) Foreground of the left figure

Fig. 6.3 Head tracking

Regarding the dataset, a group event affects a limited number of agents; in other words, the event is analyzed in certain agent subsets, which can be specified by supervised learning.

In the event detection phase, the basic technique is to search all of the subset S_t^i of the agent status set S_t, where $S_t^i \subseteq S_t$, for the event distribution in a certain subspace. This is a time-consuming approach, which has a complexity $O(2^{N_{St}})$. The efficiency of the search approach must be improved while avoiding the loss of the event.

6.3.1 *Queuing*

Starting from the location space, which is usually more important in a surveillance system, we search for a queuing event.

Given a distance $\varepsilon_k, k \in N$, we define a function $\bar{h}$ between each agent that meets the requirement

$$\begin{aligned} &\forall \ Dst(\Theta_t^i, \Theta_t^j) < \varepsilon_k \\ &\quad \bar{h}(\Theta_t^i) = \bar{h}(\Theta_t^j) \\ &\forall \ \bar{h}(\Theta_t^i) = \bar{h}(\Theta_t^j) \quad \bar{h}(\Theta_t^j) = \bar{h}(\Theta_t^l) \\ &\quad \bar{h}(\Theta_t^i) = \bar{h}(\Theta_t^l), \end{aligned}$$

(a) Four human to classify.

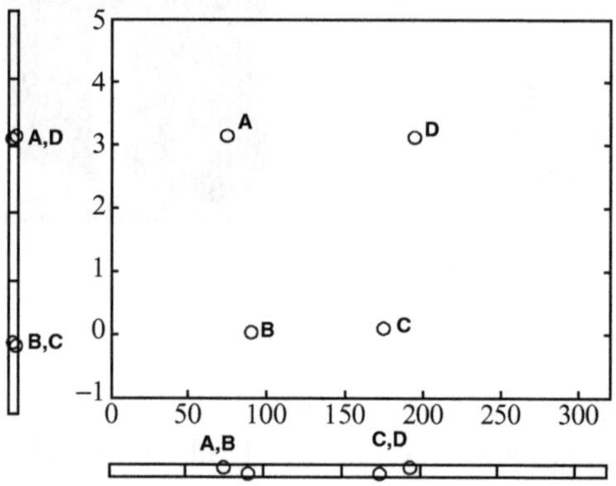

(b) Comparison of status projected on different axis.The horizontal axis represents the location of Y,the vertical axis represents the angle.

Fig. 6.4 Subspace analysis.

where $Dst()$ is the distance function in the location space.

Then, the subset S_t^{kr} can be defined as

$$S_t^{kr} = \{\Theta_t^i, \Theta_t^j \in S_t | \bar{h}(\Theta_t^i) = \bar{h}(\Theta_t^j)\},$$

so that the agent set is divided into several subsets S_t^{kr}, depending on the distance ε_k.

The distance ε_k may vary within a small range, taking into consideration the normal distance between each individual in a queue. It can also be learned from the supervised agent pair.

For each subset S_t^{kr}, suppose that the agents are distributed in a line

$$y_t^i = ax_t^i + b$$

where x_t^i, y_t^i are components of $\Theta_t^i, \Theta_t^i \in S_t^{kr}$, and the line parameters a, b can be derived from linear regression.

$$a = \frac{n\sum(x_t^i y_t^i) - \sum x_t^i \sum y_t^i}{n\sum(x_t^i)^2 - (\sum x_t^i)^2}$$

$$b = \frac{\sum y_t^i - m\sum x_t^i}{n}$$

Most importantly, the correlation coefficient r can also be estimated by

$$r = \frac{n\sum(x_t^i y_t^i) - \sum x_t^i \sum y_t^i}{\sqrt{[n\sum(x_t^i)^2 - (\sum x_t^i)^2][n\sum(y_t^i)^2 - (\sum y_t^i)^2]}},$$

which can be used to classify the agent subset as linear or nonlinear using the discriminant function learned from selected samples.

6.3.2 Gathering and Dispersing

To extend this idea to a more complex event, we try to detect agents that are gathering and dispersing in a space extended by location space and direction space.

Commonly, for each group S_t^i that is gathering or dispersing, there should be a center of motion O_t^i. The agent that belongs to this group should meet the requirement

$$\angle(O_t^i, \Theta_t^n) \pm d_t^n \leq \partial_t^i, \tag{6.1}$$

where $\angle()$ stands for the direction function from Θ_t^n to O_t^i, ∂_t^i is the tolerance, and the plus $(+)$ represents gathering and minus $(-)$ represents dispersing. Then, the manifold of the event appears in three-dimensional (3D) space, as shown in the figures below.

We can search all of the subset S_t^i that best satisfies the distribution of the function, or search all of the center O_t^i to find the agent in the manifold of Figure 6.5. This may be time consuming as it depends on the image size.

A voting method is designed to search for the center of the event, and then trace it back to the related agents. According to Equation 6.1, given an agent Θ_t^n, a family of event center O_t^i can be expected to appear in the line of Equation 6.2.

$$y = d_t^n(x - x_t^n) + y_t^n, \tag{6.2}$$

We divide the image into many bins B_t^n, which are chosen based on each agent's status Θ_t^n according to the distance from the agent to the line, compared with ∂_t^i. The prominent bins form the event center, and voting agents form the event subset.

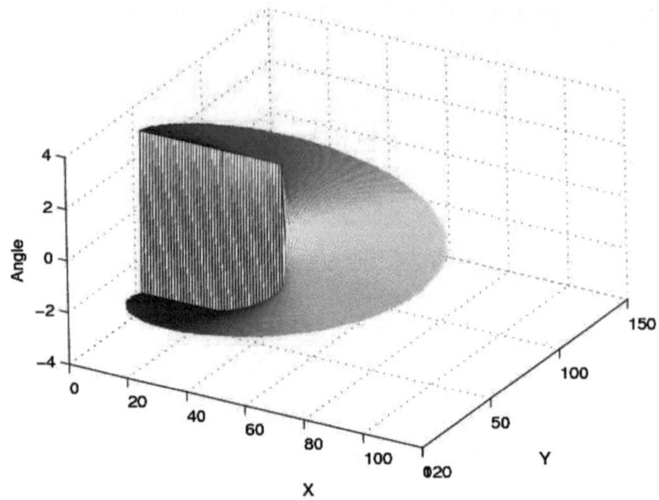

(a) Gathering

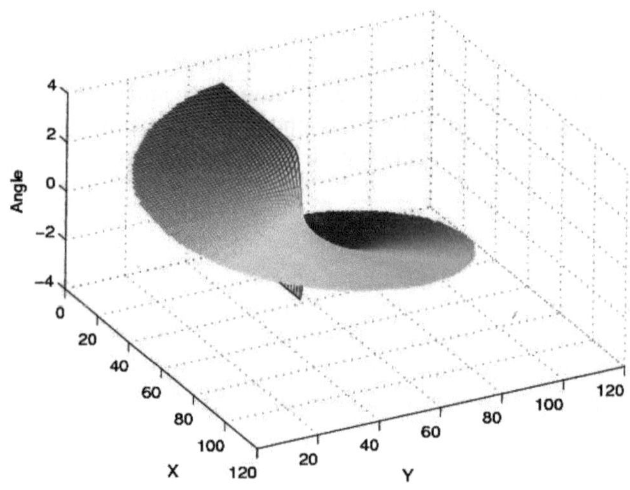

(b) Dispersing

Fig. 6.5 Manifold of gathering and dispersing event. The X and Y axes represent the x and y directions in location space, and the axis angle stands for the direction parameter of the status.

6.4 Experiments

Experiments are conducted using our method. The system is run on a 1.7 GHz PC at a speed of about 10 fps to recognize the action of each agent and detect special group events.

We have built a small library of five kinds of actions: walking, running, jumping, squatting, and stopping. Each experiment involves two sequences of different actions from four different people, and in each sequence, many action segments of about eight frames are selected as the action models. We then translate the parameter of the time segments into feature space with the Harr wavelet, which is then trained by the SVM. Then, actions of different kinds can be recognized in real time, as shown in Figure 6.6, and fewer than two frames of error are expected.

(a) Jumping

(b) Squatting

Fig. 6.6 Agent analysis of different actions.

6.4.1 Multi-Agent Queuing

In a multi-agent event, the status of each agent should first be tracked or inferred. Because of the projective effect of the camera, the locations of the agents are adjusted relative to the space of the physical ground instead of the image space. The maximum relational distance ε_k is set at about three times the width of the head, so that the agents are grouped into several subsets. For a group with more than two agents, the distribution linearity is measured with the correlation coefficient r, which is learned with the Gaussian model from some selected groups. Then the queue can be detected and drawn as shown in Figure 6.7.

Fig. 6.7 Event analysis of a queue.

6.4.2 Gathering and Dispersing

A more complex experiment is carried out to detect a gathering or dispersing event with many agents, in which both the position and direction of all agents are considered.

For the detection of a gathering event, a radial is drawn in the direction of the agent from its position, as shown in Figure 6.8. Then, a window ∂_t^i in size is slipped over the image to find the maximum value of the radial voting, which is the expected event center and identified in blue. The method can be extended to find more events at the same time.

Fig. 6.8 Event analysis of gathering and dispersing.

6.5 Conclusion

In this chapter, we address the behavior analysis of human groups. We present a novel surveillance approach that employs a multi-agent algorithm for detecting interaction for certain purposes in a classical category of surveillance tasks. Two typical group analysis situations (queuing, gathering and dispersing) are studied as examples to illustrate the implementation and efficiency of the approach. The experimental results show that this approach can be used for special surveillance tasks such as event detection.

Chapter 7
Static Analysis of Crowds: Human Counting and Distribution[1]

An important feature of crowd surveillance is the estimation of the number of persons in the monitored scene. People counting is especially important for crowd modeling as it not only gives a direct estimation of the current situation in the field of view but also contributes to other tasks including behavior analysis and crowd simulation.

In this chapter, we introduce a new learning-based method for the people counting task in crowded environments using a single camera. The main difference between this method and traditional ones is that the former uses separate blobs as the input into the people number estimator. First, the blobs are selected according to certain features after background estimation and calibration by tracking. Then, each selected blob in the scene is trained to predict the number of persons in the blob. Finally, the people number estimator is formed by combining trained subestimators according to a predefined rule. The experimental results are shown to demonstrate the functions.

7.1 Blob-based Human Counting and Distribution

Anti-terrorism is a global issue, and video surveillance has become increasingly popular in public places including banks, airports, public squares, and casinos. However, in the case of crowded environments, conventional surveillance technologies have difficulty in understanding the images captured as occlusion causes extreme problems.

Several studies have reported on this topic. Davis's group [211] was among the earliest ones to conduct extensive research into crowd monitoring using image processing. Many potential and nearly all of the open problems of crowd monitoring

[1] Portions reprinted, with permission, from Weizhong Ye, Yangsheng Xu, and Zhi Zhong, Robust People Counting in Crowded Environment, *Proceedings of the 2007 IEEE International Conference on Robotics and Biomimetics*. ©[2007] IEEE.

H. Qian et al., *Intelligent Surveillance Systems*, Intelligent Systems, Control and Automation: Science and Engineering 51, DOI 10.1007/978-94-007-1137-2_7,

have been identified. Siu [212] estimated crowd volume by counting pixels and applied a neural network for the classification, while Lin [213] estimated the number of people in crowded scenes using head-like contours. In this chapter, we use a head detection algorithm to detect human heads and treat the number of heads as the number of persons. Nikos [214] developed a crowd estimation system based on robust background modeling. While the pixel counting technique is adopted herein, the proposed method is largely based on a novel background technique. Ernesto [215] used optical flow to identify abnormal behaviors in crowd environments, and applied an HMM after the feature selection stage.

The foregoing methods can be divided into two categories: detection based and mapping based. Detection-based methods rely on a detector to count or cluster the output. The features that the detector uses may include the body, head, skin, or hair. The advantage of these methods is that they have a high level of accuracy of output for environments that are not very crowded if the detection algorithm is effective. The disadvantage is that they require a good detection algorithm and thus are not scalable for large crowds. Mapping-based methods extract features and map them onto a value. They use edge points, the background, texture, optical flow, or the fractal dimension as features. They are scalable to large crowds and give a global crowd estimation. However, it is hard to make them scene invariant, and retraining of the system is needed.

We develop a learning-based people counting system that uses separate blobs as the input into the human number classifier. In so doing, we contribute to three areas of the literature: 1) rather than applying normal camera calibration techniques, we use human tracking technology for calibration, which can be fulfilled automatically before system establishment; 2) separate blobs are used for the input into the human number classifier instead of pixel or head information, which represents an improvement over detection- and mapping-based methods; and 3) a learning-based predictor based on an SVM is applied and proven to be effective in the experiments.

The remainder of this chapter is organized as follows. Section 7.1.1 gives a detailed introduction of the system architecture. The proposed algorithm is described in detail in sections 7.1.3 and 7.1.4, including the calibration by tracking, blob-based feature input, and learning function setup. Section 7.1.5 describes an experimental study in which the proposed method is compared with other approaches. The final section presents the conclusion.

7.1.1 Overview

A number of aspects are considered in the development of the system. 1) We consider camera calibration to be important for this application and have employed a calibration-by-tracking approach that does not require human interference. The proposed camera calibration can establish the relationship between the image and world planes and map the computed image onto a more accurate space. 2) We use both background information and blob segmentation results for the classification task. These two information sources can be combined in an optimal way for crowd estimation. 3) We also deduce a density map from the people counting results, which gives a global view of the current situation and provides candidates for abnormal behavior analysis.

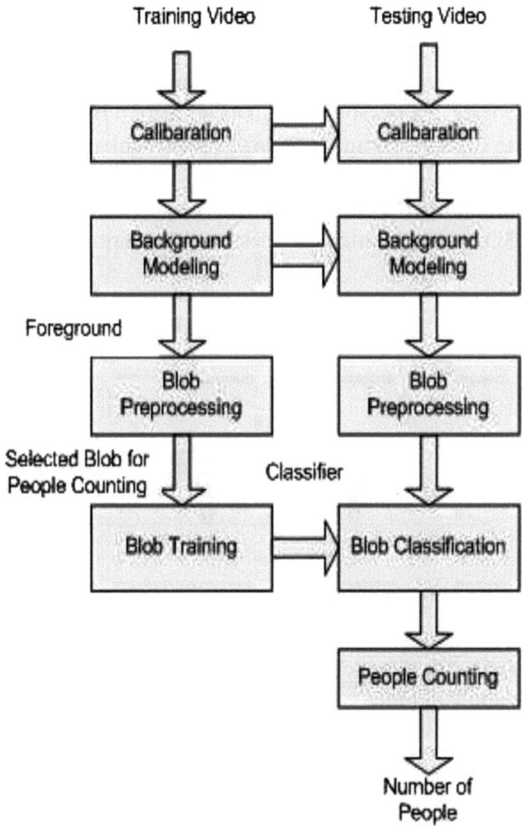

Fig. 7.1 The proposed system

The architecture of the system can be described as follows. The whole system is divided into two stages (as shown in Figure 7.1): training and testing. In the training stage, the input video is first calibrated to establish the correspondence between the image and the world plane, in which the heights of persons are supposed to be equal. Thus, the world plane can be transformed into a 2D one. We use a calibration-by-tracking method to fulfill the task. Then, we apply an improved adaptive Gaussian mixture model (GMM) [216] for background modeling. After the background is obtained, a rule-based algorithm is used to select blobs for training to identify the total number of persons in each blob. Finally, the number of persons in each blob is combined to get the total number of people in the scene.

7.1.2 Preprocessing

Camera calibration is important in video surveillance applications. Even if the camera is well placed in a location and the image quality and size of a person are acceptable, distortions exist. Also, the distance between the camera and a person is not the same, and persons of different size appear in the image. In the proposed system, we apply a calibration-by-tracking approach. The calibration is performed in the following way. First, we have a person walk in as straight a line as possible towards the camera. At the same time, the height and area of the blob of this person in the image are recorded. Then, we establish the relationship between the height of the blob in the image and the person's real height. In this way, the 2D image information can be mapped onto a 3D map, in which all persons are supposed to be the same height.

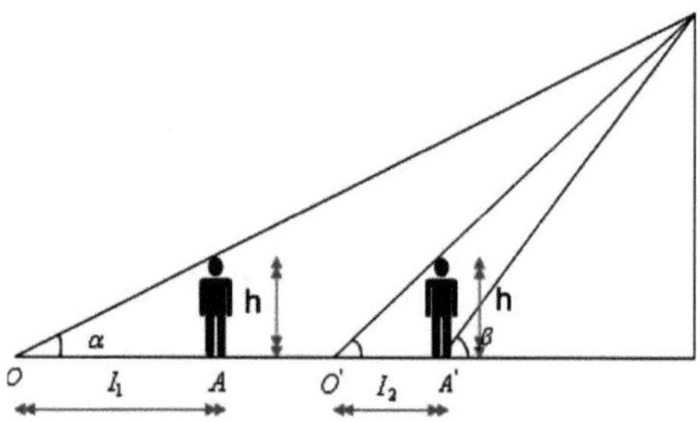

Fig. 7.2 Calibration by tracking approach.

In Figure 7.2, I_1 is the distance between point O and point A, while I_2 is the distance between point O' and A'. From the figure, we can easily derive

$$I_1/I_2 = tan(\alpha)/tan(\beta); \tag{7.1}$$

When a new image is entered, it is transferred into a new space according to the relationship between the camera and real world.

7.1.3 Input Selection

With the achievement of the calibration, the correspondence between the image and world planes is well established. In this chapter, we do not use all pixel information as the input into the people counting system but use separate blobs for people counting. The reason for doing so is threefold: 1) pixel information inevitably involves system noise such as human distortion, shadows, and so forth; 2) the separate blob approach can be used to handle problems when only parts of people are in the scene, whereas the all-pixel method cannot; and 3) if the blobs are separated, then we can get the location of each blob. This information is also very meaningful for tasks such as location-aware surveillance and behavior analysis in the crowd modeling process.

The task of blob selection for system training involves selecting the blobs that contain people. Thus, we give up blobs that are relatively small as they are very likely to be noise caused by background modeling and other processes. We also develop methods to handle such cases where only part of a person is inside the scene. The limited pixels are counted as a single person or several persons in a group. People are always blocked by each other when they draw near to one other. These cases are partly solved by training the blob according to the location-related pixel information in the blob, such as the center of the blob, the size of the ex-rectangle, the distribution of the pixels in the blob, and the shape of the blob itself.

The main difference between the proposed method and others is the input selection for the classifier. Here, we use segmented blobs as the input into the people counting classifier. The input selection is rule based, and the details can be seen in Figure 7.3.

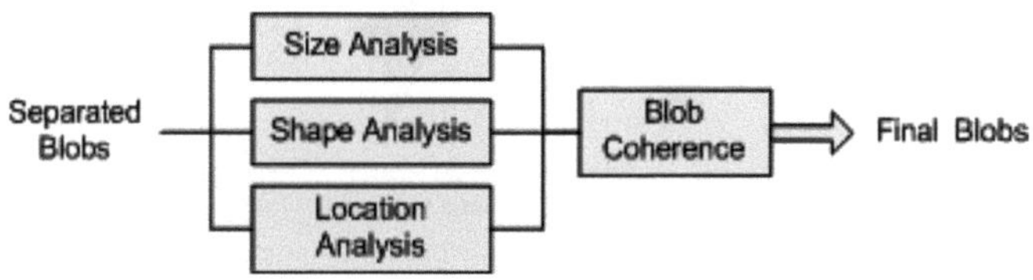

Fig. 7.3 Blob selection procedure.

Size Analysis

The size analysis module is based on the assumption that the size of persons is confined within a reasonable range. Raw images contain many small blobs, which may be small parts of the human body. Noise may include small parts of trees and rivers and some small blobs produced by background modeling. If these blobs are used as input into the classifier, then the image processing will be computationally heavy. Thus, we omit blobs that are smaller than a predefined number. The predefined number can be learned or given manually. In this way, we can delete much noise, such as that noted above.

Shape Analysis

The shape analysis module is based on the assumption that the shape of persons is similar to an ellipse or related shape. Thus, other objects such as dogs, cats, and buses can generally be filtered. Taking a bus as an example, when it enters the field of view, the result obtained using traditional methods indicates a significant number of people. In addition, in the shape analysis module, we use vertex points as the input into the classifier, which shows the distribution of the pixels in a blob.

Location Analysis

The location analysis module can handle different cases, i.e., when people are near or far from the camera. It is also effective for cases where someone has just entered the environment, and only part of the person can be seen (such as the head or the leg). If the segmented blob is at the edge of the monitored scene, then the module judges the location for special processing.

Blob Coherence
Some cases still happen when one person is segmented into different parts. If the number of pixels in a blob falls within a predefined pixel number, then they will be judged to represent more than one person. The number of persons counted may then be greater, which will result in inaccurate counting. The blob coherence module combines neighboring blobs, which are very close to each other. These blobs are likely to originate from a single person. If the blobs are from different persons, then the system will regard them to be multiple persons, which will not introduce errors. Typical cases are shown in Figure 7.4.

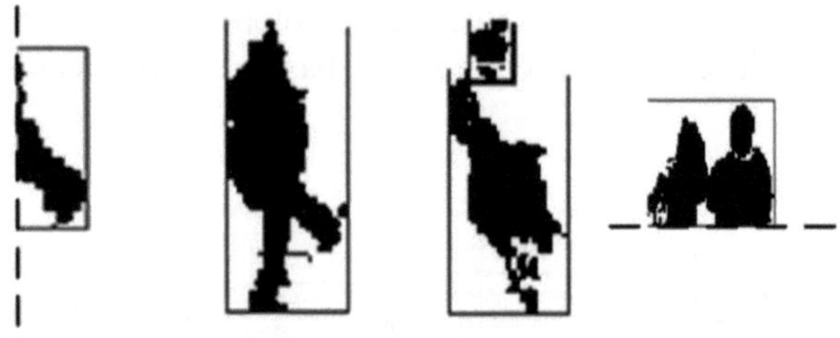

Fig. 7.4 Different cases of blob selection

7.1.4 Blob Learning

After the blobs are selected as the input into the classifier, they are learned to predict the number of persons. Thus, the identification problem of the total number of people can be treated as a multiclassification problem. Here, an SVM is adopted as the classifier of the human number predictor. The abovementioned tool has demonstrated satisfying results in gait recognition for a multiclassification task in intelligent shoe applications [217], which are very similar to our experiments.

The blob learning method here is to learn every preprocessed blob to identify how many persons are in the blob. Before blob learning, the system needs to confirm whether the input blob represents multiple persons or only one part of a person. The motivation for using a learning method instead of a traditional pixel counting one is that the number of persons in one blob cannot simply be deduced from the number of pixels but is related to the shape, size, and location of the blob.

Input Parameters
The input parameters that we select for the initial experiments are the location of the blob in the image, including the x- and y-values of the center of gravity of the blob, the location of a series of vertexes along the convex hull of the blob, and most importantly, the number of pixels in the corresponding blob. The vertexes of the blob will be normalized for the input into the classifier. The x-value of the center of the blob is G_x and its y-value is G_y. The vertexes of the blob can be written as $T = [t(1), t(2), t(3)\ , \dots , t(l)]$. The number of pixels in the blob is N. The normalized

vertexes of the blob can be written as $T = [t(x_1), t(x_2)\ , t(x_3)\ , \dots , t(x_m)]$, where $t(x_i)$ is selected from $t(i)$. The input parameters can finally be described as follows:

$$I = [G_x, G_y, N, t(x_1), t(x_2), t(x_3), \dots, t(x_m)] \tag{7.2}$$

Output Parameters

We now have a multiclassification problem. The output of the classifier is the number of persons in a blob. Here, we suppose that we have 10 possible results. If the number of persons is larger than 10, then we use a simple pixel counting method to fulfill the task, although these situations rarely happen.

Learning Tools

SVMs are popular learning tools for handling nonlinear data. Here, we choose an SVM as the learning tool to train and test the data.

SVMs perform pattern recognition between two classes by finding a surface at the maximal distance from the closest points in the training sets. We start with a set of training points x_i that belongs to R^n $(i = 1, 2, \dots, N)$, where each point x_i belongs to one of two classes identified by the label y_i, which belongs to $\{-1, 1\}$. Assuming that the data are linearly separable, the goal of maximum margin classification is to separate the two classed by a hyperplane such that the distance to the support vectors is maximized. The hyperplane can be described as follows:

$$f(x) = \sum \alpha_i * y_i * x + b \tag{7.3}$$

In the experiments, we solve a p-class problem with SVMs. There are basically two approaches: one-vs.-all and pairwise. We adopt the former approach. In this method, p SVMs are trained and each of the SVMs separates a single class from all remaining classes.

7.1.5 Experiments

In our experiments, we compare our method with the all-pixel one and demonstrate that ours is more robust and accurate. We also conduct density map experiments using the results of people counting.

For the experimental setup, we use a digital camera to collect the data, which are recorded in a railway station in Hong Kong. The resolution of the video is 320*240 pixels and the duration of the video is 30 minutes. The objects in the video that we record include only people, not cars or animals. In the 30-minute video, 10 minutes are used for training and 20 minutes for testing. Camera calibration is done independently.

The process of training data collection is described as follows. We design an interface for users to input the information, including whether the blobs contain persons. If the answer is positive, then we continue to count the number of people in the blobs. We collect approximately 1500 blobs of different size and location for training.

Blob Selection Result

One set of blob selection results can be seen in the first column of the figure below. The left image is the raw image with five persons: only the head of one person is in the field of view, two persons are walking separately, and the other two are walking very close to each other, so that they are detected as one single blob. The right image is the resultant image after blob selection. From Figure 7.5, we can see that the right blobs are selected and the noise is removed.

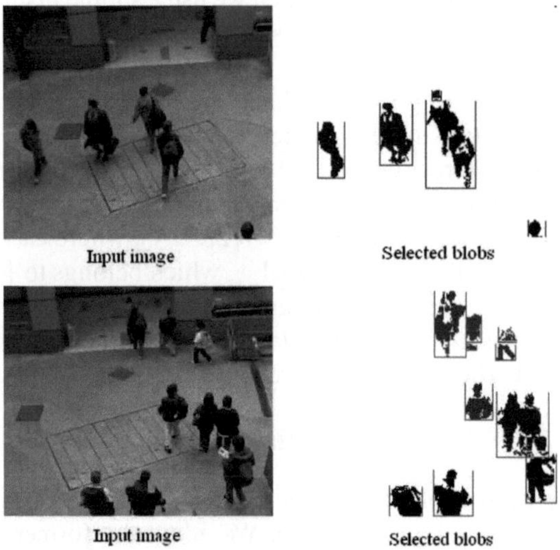

Fig. 7.5 Blob selection result.

People Counting Result

A people counting experiment is carried out by comparing the result obtained using our method with that using the all-pixel method. In Figure 7.6, *m* is the result of the former method, *l* is the result of the latter one, and *n* is the actual number of persons.

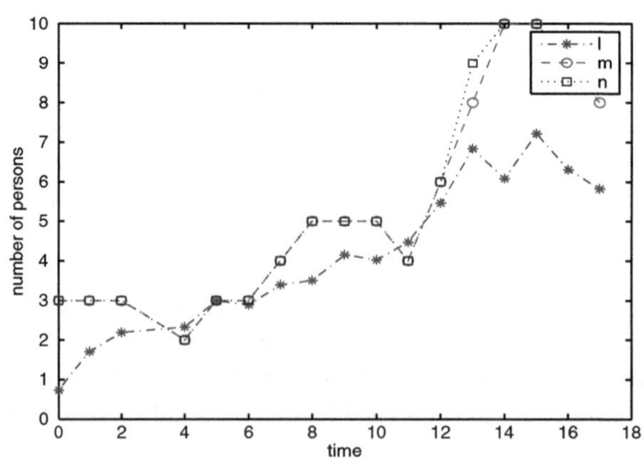

Fig. 7.6 Comparison of different methods.

7.1.6 Conclusion

In this section, we present a robust people counting system for crowded environments. The proposed system is unique in that it adopts a calibration-by-tracking method for person calibration before people counting starts. Then, most importantly, separate blobs are used as the input into the person number classifier, which improves the accuracy of the person counting system. In addition, a density map is produced in real time using the resultant data, which can potentially be used as candidates for other systems such as the behavior analysis module and crowd simulation system.

7.2 Feature-based Human Counting and Distribution

In the case of crowded environments, conventional surveillance technologies have difficulty in understanding the images captured as the incidence of occlusion is high. In this chapter, we present a methodology for modeling and monitoring crowds that is based on the theory of learning from demonstration. Teaching a surveillance system several crowd density levels equips it with the ability to draw a density distribution map. A dynamic velocity field is also extracted for teaching the system. Thereafter, learning technology based on appropriate features enables the system to classify crowd motions and behaviors. The goal of our research is to generate a stable and accurate density map for the input video frames, and to detect abnormal situations based on density distribution, such as overcrowdedness or emptiness.

Crowd control and management are very important tasks in public places. Historically, many disasters have occurred because of the loss of control over crowd distribution and movement. Examples include the Heysel stadium tragedy in 1985, the Hillsborough tragedy in 1989, and, more recently, the incident in 2004 when many people lost their lives at the Festival of Lanterns in Miyun, Beijing. It is extremely important to monitor the crowdedness of different locations in tourist areas, such as mountain parks, for the sake of safety. To prevent the problems caused by large crowds and to be able to respond quickly to emergent situations, the proper control and management of crowds is vital. Generally speaking, the problem of crowd control can be divided into two main aspects: real-time monitoring of crowd status and the design of the environment based on models of crowd behavior.

Crowd density is one of the basic descriptions of crowd status. With density estimation, it is easy to determine the crowd distribution and identify abnormal changes in a crowd over time. However, crowd density estimation is difficult because many occlusions occur in crowded environments, and generally only parts of people's bodies can be observed. Also, as the number of people increases, identification becomes harder. Therefore, many approaches that are based on the detection of body parts, such as the head, body, or legs, may not work properly. Another difficulty lies in the generation of a density map. For most applications, a pixel-level density map is unnecessary; therefore, we must find another way to partition the image cells to measure density. Because of the nonuniformity of 2D projections of 3D objects, the size of the cells will be different in different parts of the image, and thus a reasonable partition scheme needs to be developed.

The goal of our system is to generate a stable and accurate density map for the input video frames and to detect some abnormal situations based on density distribution, such as overcrowdedness or emptiness.

7.2.1 Overview

In this section, we present an automatic surveillance system that can generate a density map with multiresolution cells and calculate the density distribution of the image using a texture analysis technique. The system has two main parts: density map calculation (Figure 7.7) and abnormal density detection (Figure 7.8).

There are two phases in the flowchart of the density map calculation (see Figure 7.7): training and testing.

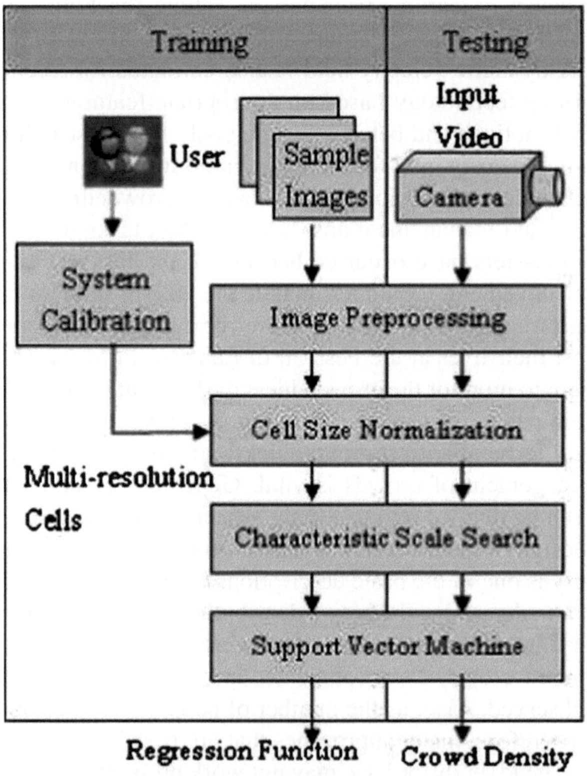

Fig. 7.7 Block diagram of the density calculation process.

The objective of the training phase is to construct a classifier that can be used to calculate the density. First, a series of multiresolution cells is generated by an initial calibration process that is based on a perspective projection model. To achieve a uniform texture representation, the size of different cells is normalized to a certain value. Then, using the texture analysis approach, a texture feature vector that contains sixteen parameters is calculated for each image cell. Because there is an inevitable change in scale between image cells, obtaining a characteristic texture feature representation requires the creation of a group of image cells of different scale using the Gauss-Laplacian kernel function. By searching for the maximum point in the Harris-Laplacian space, the characteristic feature vector for each cell can be calculated. Once it has been achieved, all vectors and the corresponding density estimations of the sample images are fed into an SVM system.

Our system makes three main contributions. 1) By performing texture analysis of both the local and global areas in the image, the system can construct a multiresolution density map that describes crowd distribution. 2) To improve the stability of the crowd density estimation, a search algorithm is applied in both the image and the scale space. 3) The density estimation is implemented using the approach of learning by demonstration.

As there are a large number of instances of abnormal crowds gathering in public places, we also implement an abnormal density detection system, as shown in Figure 7.8. We first create some abnormal simulated density data using Matlab, and use the SVM approach to detect abnormal situations in real videos taken from Tiananmen Square in Beijing.

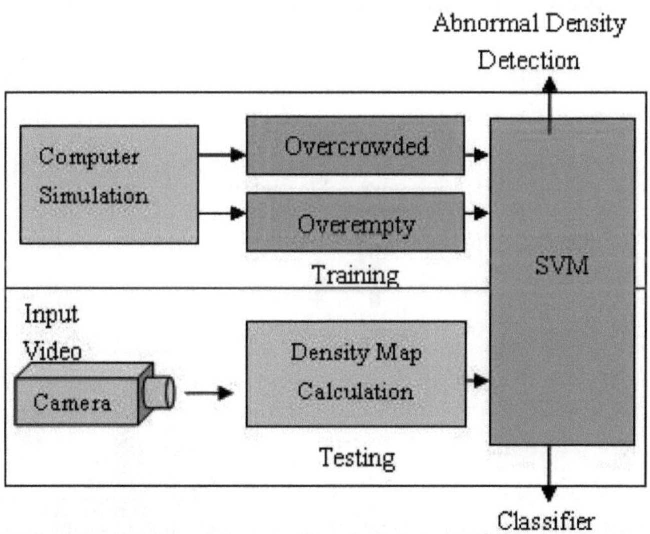

Fig. 7.8 Block diagram of the abnormal density detection process.

7.2.2 Initial Calibration

The objective of the system is to build a density map. Ideally, the density map should be a bending surface to indicate the crowd distribution in the image. The most common method is to define a rectangle whose center is located in a certain pixel in the image, and then calculate the density in this area pixel by pixel. However, this method is time consuming, and building a pixel-level density map is unnecessary for most applications. An alternative solution is to divide the area of interest in the image into a series of cells, and calculate the corresponding density in each cell.

Another problem is caused by the 2D-3D projection. As Figure 7.9 illustrates, objects of the same size are smaller in the image when they are further away from the camera, and thus the cell size should be different to achieve the same texture measurement. A multiresolution density cell model is therefore considered. The cells may have small overlapping areas, because we want the covered image area to be as large as possible to satisfy the perspective projection model, which is discussed in the next section.

The initial calibration involves the procedure of building a multiresolution cell model based on geometric analysis.

7.2.2.1 Multiresolution Density Cells with a Perspective Projection Model

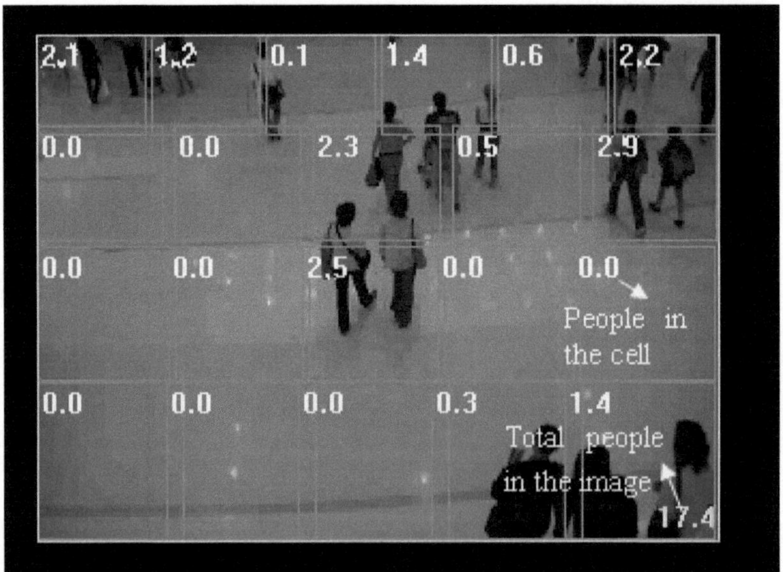

Fig. 7.9 Multi-resolution density cells indicating the estimated number of people in each cell and the entire area (In this case, the estimated number of people is 17.4).

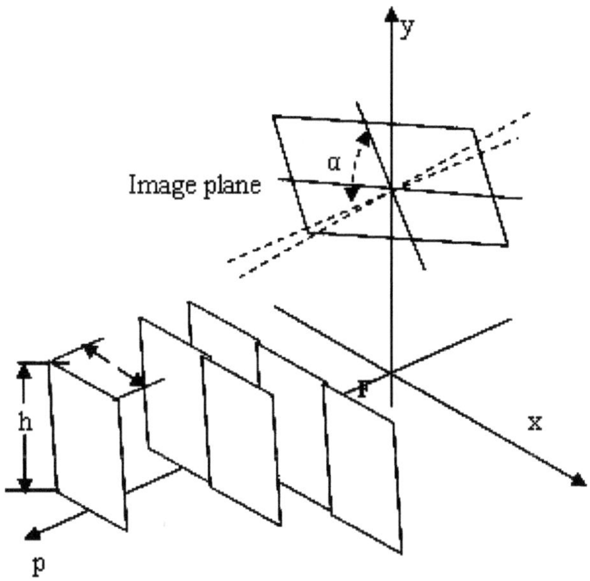

Fig. 7.10 Model of density cells projection.

We present a method that is based on a perspective projection model to generate a series of multiresolution density cells. There are four conditions assumed in the model: that all of the people are of a similar size, all of the people are located in the horizontal plane, the camera only rotates along a vertical axis, and the image center is the optical center of the camera.

Suppose that there are a number of parallel rectangles representing density cells in a 3D space. These rectangles are not equidistant but are perpendicular to the horizontal plane. The height of the rectangles is h and the width is w, as Figure 7.10 illustrates. Under the condition of the perspective projection, the projections of these rectangles are the density cells in the image.

As Figure 7.11 shows, the camera's rotation angle is $\alpha(\pi/2 < \alpha < 0)$, its focal length is f, and H is the distance from the camera focus to the horizontal plane. We assume that the rotation angle α is known, but it can easily be calculated by the vanishing point method [213] or read directly from a measuring instrument.

To ensure the uniformity of the texture measurements, the user needs to define the area of interest in the input image, together with the height and width (in pixels) of the person nearest to the camera in the image, and the scalar of the cell height relative to the person's height.

The projections of these lines change with the camera rotation angle α. The distribution of the projections can be calculated by

$$\theta_i = \arctan(\frac{\overline{op_i}}{f}\mu) \tag{7.4}$$

$$\theta_{i+1} = \arctan(\frac{\overline{op_{i+1}}}{f}\mu) \tag{7.5}$$

where $\overline{op_i}$ and $\overline{op_{i+1}}$ are the lengths in pixels, and μ is the real length of a pixel in the image plane.

According to the geometric analysis in Figure 7.11, four equations can be applied:

$$\frac{\overline{p_i p_{i+1}}}{\sin(\alpha-\theta_{i+1})} = \frac{s_1}{\sin(\pi/2+\theta_{i+1})} \tag{7.6}$$

$$\frac{s_1}{h} = \frac{f/\cos\theta_i}{H/\cos(\alpha-\theta_i)} \tag{7.7}$$

$$\frac{s_2}{h} = \frac{f/\cos\theta_{i+1}}{H/\cos(\alpha-\theta_{i+1})} \tag{7.8}$$

Substituting Equations (7.7) and (7.8) into Equation (7.6), we get

$$s_2 = \frac{\overline{p_i p_{i+1}}\cos\theta_i\cos(\alpha-\theta_{i+1})}{\sin(\alpha-\theta_{i+1})\cos(\alpha-\theta_i)} \tag{7.9}$$

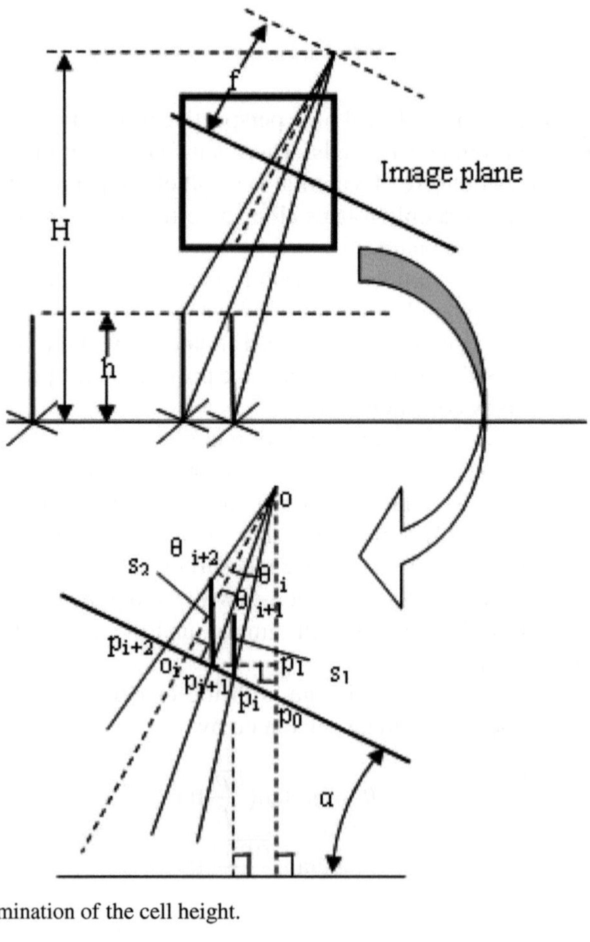

Fig. 7.11 Determination of the cell height.

In $\triangle p_0 p_{i+2} o$,

$$\overline{p_o o} = \frac{f \sin(\alpha - \theta_{i+1})}{\cos\theta_{i+1}\sin(\pi/2 - \alpha)} \tag{7.10}$$

$$\overline{p_o p_{i+1}} = \frac{f}{\cos(\pi/2 - \alpha)} \tag{7.11}$$

Using similar triangles, the following equation can be found.

$$\frac{\overline{p_{i+1} p_{i+2}}}{\overline{p_{i+1} p_{i+2}} + \overline{p_o o}} = \frac{s_2}{\overline{p_o p_{i+1}}} \tag{7.12}$$

By substituting Equations (7.9), (7.10), and (7.11) into Equation (7.12), the equation that determines the length of $\overline{p_{i+1} p_{i+2}}$ can be derived. We define three variables to simplify the expression.

$$I_0 = f\cos\theta_i \sin\alpha \cos(\alpha - \theta_{i+1})$$
$$I_1 = f\cos\theta_{i+1}\cos\alpha\cos(\alpha - \theta_i) sin(\alpha - \theta_{i+1})$$
$$I_2 = \cos\theta_i \cos(\alpha - \theta_{i+1})\sin\alpha\cos\alpha\cos\theta_{i+1}$$

The equation is an iterative equation:

$$\overline{p_{i+1} p_{i+2}} = \frac{\overline{p_i p_{i+1}} I_0}{I_1 - \overline{p_i p_{i+1}} I_2} \tag{7.13}$$

There are three cases, as shown in the following.

1. If $\overrightarrow{op_i} > 0$ and $\overrightarrow{op_{i+1}} > 0$, then the iterative equation is Equation (7.13).
2. If $\overrightarrow{op_i} > 0$ and $\overrightarrow{op_{i+1}} < 0$, then the iterative equation should be Equation (7.13) with θ_{i+1} replaced by $-\theta_{i+1}$.
3. If $\overrightarrow{op_i} < 0$ and $\overrightarrow{op_{i+1}} < 0$, then the iterative equation should be Equation (7.13) with θ_i replaced by $-\theta_i$, and θ_{i+1} replaced by $-\theta_{i+1}$.

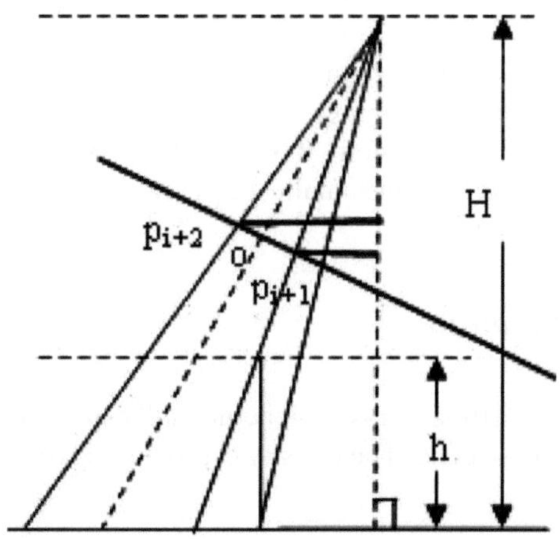

Fig. 7.12 Determination of the cell width.

After the distance of the projection of those parallel lines has been calculated, the area of interest in the image is divided into several rows of different height. Each row is then divided to make several cells.

As Figure 7.12 shows, by letting w_i, $w_i + 1$ denote the cell width in rows i and $i+1$, the cell sizes can be drawn by the following equations.

$$\frac{f\cos(\alpha-\theta_{i+1})}{(H-h)\cos(\theta_{i+1})} = \frac{w_i}{w} \tag{7.14}$$

$$\frac{f\cos(\alpha-\theta_{i+2})}{(H-h)\cos(\theta_{i+2})} = \frac{w_{i+1}}{w} \tag{7.15}$$

By substituting Equation (7.15) into Equation (7.14), w_{i+1} can be calculated by

$$w_{i+1} = w_i \frac{\cos(\alpha-\theta_{i+2})\cos\theta_{i+1}}{\cos(\alpha-\theta_{i+1})\cos\theta_{i+2}} \tag{7.16}$$

Once $\overline{p_{i+1}p_{i+2}}$ has been calculated, then θ_{i+2} can be calculated by

$$\theta_{i+2} = \arctan(\frac{\overline{op_{i+2}}}{f}\mu) \tag{7.17}$$

Given the cell width in the first row, the width in the other rows can be calculated iteratively. It should be noted that we assume that the camera rotates only around the vertical axis, and thus all the cell widths in a row are the same. Finally, all of the density cells are generated by geometric analysis based on the perspective projection model.

7.2.2.2 Normalization of Density Cells

We use the gray level dependence matrix (GLDM) to build the feature vector, which means that some of the features are sensitive to the cell size. It is thus necessary to normalize the cell size to achieve a uniform representation of the texture measurements. The normalized size can be neither too small nor too large. Overly small cells cannot preserve enough texture information, whereas overly large ones usually induce interpolation of the small image cells, which introduces errors into the texture feature calculation. In our system, the normalized size is set experimentally to a value that is a little larger than the smallest cell size.

7.2.3 Density Estimation

The goal of the density estimation is to generate a density map based on the aforementioned multiresolution cells. The density estimation involves the following four steps.

1. Image preprocessing
2. Feature extraction

3. Searching for the characteristic scale
4. System training

To decrease the influence of a change in illumination, histogram equalization is used for the image preprocessing. The features based on the GLDM method are extracted in each normalized cell. To achieve more uniform texture measurement, we apply the technique of searching for extrema in the scale space of the image cells, and use the features of the characteristic scale to form the feature vectors. It can be regarded as a procedure of feature refining. Finally, the derived features are used in an SVM-based training system.

7.2.3.1 Feature Extraction

There are many methods for texture feature extraction, as has been stated. Structural methods are the most easy and efficient to implement. However, spectrum analysis methods can extract features with prominent physical meaning, and statistical methods are easy to apply and the extracted features are more stable against disturbances. Statistical methods are thus the preferred choice for this research.

The GLDM method proposed by Haralick [29] is a classical statistical method for texture feature extraction. In essence, it is a statistical method that is based on the estimation of the second-order joint conditional probability density function, which is the probability of a pair of grey levels (i, j) occurring in a pair of pixels in an image where the pixel pairs are separated by a distance d in the direction θ. The estimation uses 2D histograms, which can be written in a matrix known as a concurrent matrix, or GLDM.

The concurrent matrix can be calculated given a pair of parameters (d, θ). Normally, one pixel distance and four directions $(0^o, 45^o, 90^o, 135^o)$ are used.

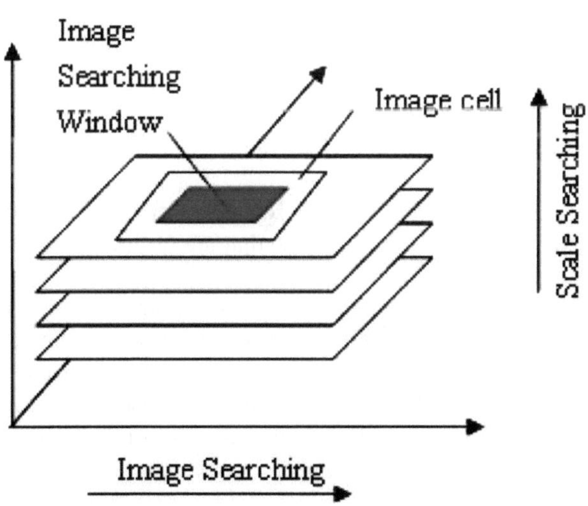

Fig. 7.13 Searching for the characteristic scale of the image in each cell.

For a pair of parameters (d,θ), four indicators with certain physical meaning can be calculated based on the corresponding GLDM as follows.

Contrast:

$$S_c(d,\theta) = \sum_{i=0}^{L-1}\sum_{j=0}^{L-1}(i-j)^2 f(i,j|d,\theta)$$

Homogeneity:

$$S_h(d,\theta) = \sum_{i=0}^{L-1}\sum_{j=0}^{L-1}\frac{f(i,j|d,\theta)}{(i-j)^2+1}$$

Energy:

$$S_g(d,\theta) = \sum_{i=0}^{L-1}\sum_{j=0}^{L-1} f(i,j|d,\theta)^2$$

Entropy:

$$S_p(d,\theta) = \sum_{i=0}^{L-1}\sum_{j=0}^{L-1} f(i,j|d,\theta)\log(f(i,j|d,\theta))$$

where $f(i,j|d,\theta)$ is the joint conditional probability density function, and L is the number of grey levels. Normally, it is unnecessary to use all 256 levels, and in our system, 32 levels are used for better computational performance.

The feature vector consists of 16 components, including four indicators along four directions. For each image cell, the feature vector is extracted for the next search stage.

7.2.3.2 Searching for the Characteristic Scale

Although the use of multiresolution cells reduces the nonuniformity of the texture measurements, some cells, especially those that are far away from the optical axis, still exhibit instability.

To arrive at a more robust representation of the texture feature of the image cells, we present a method in which a 2D search is performed in the scale and feature spaces. The method is inspired by the scale invariant feature transform (SIFT) algorithm in [96]. The search algorithm includes two phases. First, we search for the extrema of the Harris response in the image space. Second, the scale space is built, and we search for the extrema in the scale space. The algorithm is illustrated in Figure 7.13.

Image Space Searching

The Harris method [97] is often used to detect corners. In essence, its image derivative matrix reflects the degree of local density changes in the window. The Harris response is the function of the two eigenvalues of the image derivative matrix. A corner point means two large eigenvalues, and an edge means one large and one small eigenvalue.

The same idea can be applied in the measurement of image texture. Let $(s_1, s_2, \ldots, s_{16})$ denote the GLDM feature vector. We can then define a texture matrix M using the equation

$$\mathbf{M} = \begin{pmatrix} \sum_{i=1}^{L} s_i^2 & \sum_{i=1}^{L}\sum_{j=1}^{L} s_i s_j \\ \sum_{i=1}^{L}\sum_{j=1}^{L} s_i s_j & \sum_{j=1}^{L} s_j^2 \end{pmatrix} \tag{7.18}$$

and can define the Harris response as

$$R = det(\mathbf{M}) - K(trace(\mathbf{M}))^2 \tag{7.19}$$

where K is an empirical constant. In each image cell, we define a search window that is centered in the cell. To achieve better performance, we determine the search window size by the following equation:

$$L_s = \min(L_{cell}/5, 5) \tag{7.20}$$

where L_{cell} is the size of the image cell. We then search for the largest local maximum Harris response in the search window. If there is no local maximum, then the largest response is used.

Scale Space Searching

After the Harris search, the GLDM feature vector is calculated for each image cell and a multiscale space is built. The scale space representation is a set of images presented with different levels of resolution that are created by convolution with the Gaussian-Laplacian kernel.

$$J(X, s) = L(s) * I(X) \tag{7.21}$$

where $X = (x, y)$ is a pixel in the image, and s is the parameter of the Gaussian-Laplacian kernel that controls the scaling transformation. The Gaussian-Laplacian kernel function is expressed as

$$L(s) = s^2(G_{xx}(x, y, s) + G_{yy}(x, y, s)) \tag{7.22}$$

A set of images with different scales for each cell is built, and the normalized image cell is regarded as the finest resolution image. The Harris response is then calculated again to search for extrema in the scale space. If there are none, then the normalized scale is considered to be the characteristic scale.

7.2.3.3 System Training

The SVM method [42] is again applied to solving the nonlinear regression estimation problem. First, we calculate the texture feature vector in the characteristic scale for each cell in all of the sample images. The crowd density for each cell is then estimated by a human expert. The problem involves establishing the relationship between the input vector and the output density. This is a typical regression problem, which can be solved by the SVM method. The expression of the estimated function is

$$f(\overrightarrow{x}) = \sum_{i=1}^{L}(\alpha - \alpha^*)K(\overrightarrow{x_i}, \overrightarrow{x}) + b \tag{7.23}$$

where α, α^* are the Lagrangian multipliers, and $\overline{x_i}$ are support vectors. We apply a polynomial kernel $K(\overrightarrow{x_i}, \overrightarrow{x}) = (\overrightarrow{x_i} \cdot \overrightarrow{x} + 1)^d$, where the kernel parameter $d = 2$ and the adjustment parameter $C = 12$ in the loss function.

7.2.4 Detection of an Abnormal Density Distribution

We have developed an abnormal crowd density detection system that can detect some typical abnormal density distributions by machine learning. The system is based on the density map generated by the aforementioned method. The system flowchart is shown in Figure 7.8, and again involves two stages: training and testing.

Once the density map has been calculated, the crowd density distribution is known. Crowd density changes can be observed directly from the density map. In the real world, and especially in large public places, a change in density may indicate potential danger or an emergency, and it is therefore very useful for detecting abnormal distributions in crowd density.

7.2.4.1 Training Data Created by Computer Simulation

After initial calibration, the size and position of the density cells are determined so that a vector containing all of the density data can be built. From the changes in the vector's components, we can track the variations in the density distribution.

We record some density vectors for abnormal situations, and use them as sample data. We can then detect similar situations using the machine learning method.

Because of the lack of real density data on abnormal situations, a set of training data are created by computer simulation. Here, we consider only two abnormal cases: overcrowdedness and emptiness in a local area of the region under surveillance. About 180 scenes are created, including positive (overcrowded) and negative (normal) samples.

Fig. 7.14 Crowd density estimation of frames captured from a testing video in Hong Kong.

7.2.4.2 System Training and Testing

The objective of the training system is to divide the images into two classes, abnormal and normal density distribution, according to the calculated crowd density features. As noted, many types of learning systems that can be used for such a binary classification problem, including SVMs, RBFNs, the nearest neighbor algorithm, and Fisher's linear discriminant. The SVM method is chosen for the classifier training because it has stronger theory interpretation and better generalization

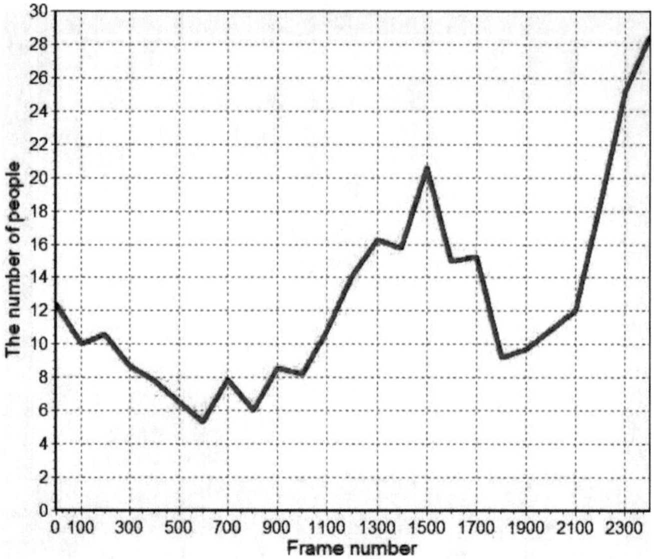

Fig. 7.15 Estimation of the number of people over time.

performance than the other neural networks. SVMs also have such advantages as very few tunable parameters and the use of structural risk minimization.

To train the system, two steps must be followed. First, a database of positive and negative samples is created by computer simulation. Second, the samples are fed into the SVM training system to construct the classifier. A polynomial kernel is used again, where the kernel parameter $d = 2$, and the adjustment parameter C in the loss function is set to 10.

7.2.5 Experiment Results

The proposed system is tested with real videos taken in Hong Kong and Beijing. For the density map calculation, several video clips taken in Hong Kong are used. About 70 images extracted from the video are used as samples. The resolution of the images is 320×240 pixels, and the whole image area is divided into 21 cells in the calibration stage. The crowd number for each cell in the sample image is counted by an expert for initial training and evaluation. The maximum estimation error for each cell is below 5%, and the total crowd number estimation error is below 12%.

Figure 7.14 shows some frames from the testing video, in which the estimated number of people in local cells and the total number of people are displayed. Figure 7.15 shows the change in the estimated total number of people with respect

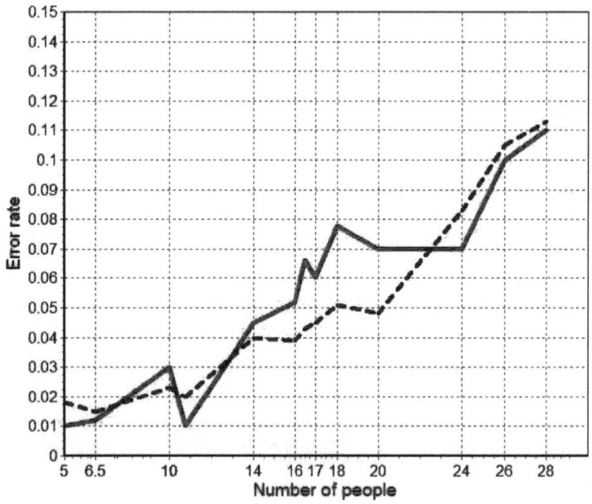

Fig. 7.16 Comparison of the error rate with respect to the number of people with (dotted line) and without (solid line) applying the searching algorithm.

to the frame number, and clearly shows a change in crowd density in the image sequence. The cells may have small overlapping areas because we want the covered image area to be as large as possible. The position of the density cells in each row may be a little different from what Figure 7.10 shows, because each cell shifts along the line parallel to the x-axis to cover the whole image area in the best way possible.

To examine the effectiveness of the searching algorithm, we compare the error rate curve of the density estimation with respect to the number of people by applying and not applying the search algorithm, as Figure 7.16 shows. The results reveal that the error rate becomes larger as the number of people increases, especially when there is much overlap in the local area, but most of the rates are below 10%. The search algorithm gives better results when the scene contains a moderate number of people.

Several video clips taken in Tiananmen Square, Beijing, are used for the abnormal density detection experiment. Figure 7.17 shows people gathering at the right corner in the image sequence, and the top and middle images show the change in crowd distribution. The bottom image is a snapshot of the system giving an alert that a situation of local overcrowding has occurred.

The results indicate that our processing algorithm is effectively able to handle foggy outdoor environments crowded with humans that appear small in the camera images, as well as indoor environments with significant body occlusion problems.

Fig. 7.17 The overcrowded situation detected from a video of Tian'anmen Square (the numbers are the estimated number of people in each cell).

7.2.6 Conclusion

We present an approach for calculating a crowd density map from video frames. In the initial calibration stage, a set of multiresolution density cells is created based on a perspective projection model. The GLDM feature vector is extracted for each cell. A scale space is then built and the characteristic scale obtained by searching in the Harris-Laplacian space. Finally, the feature vectors are fed into an SVM training system to solve the nonlinear regression problem. An abnormal density detection system based on multiresolution density cells is also introduced.

Chapter 8
Dynamic Analysis of Crowd Behavior[1, 2, 3, 4]

8.1 Behavior Analysis of Individuals in Crowds

In many cases, we cannot obtain the background image easily, and if the occlusion is serious, then it is difficult to segment masses of blobs into single blobs. We therefore use optical flow to detect abnormal behavior in crowded environments [57], [58].

The velocity of each pixel is calculated using optical flow and then filtered using a 3×3 template. A predetermined threshold is used to obtain the binary images. If the value of the filtered velocity is larger than the threshold, then we set the gray value of this pixel to 255; otherwise, it is set to 0.

The image on the screen is distorted because of the perspective rule, and thus a camera model [164] is used and compensation algorithms are employed so that the length and speed are roughly independent of the depth of a blob from the camera.

After the binary images are obtained, ordinary methods are used to detect abnormal behavior in a very crowded environment.

In this experiment, we aim to detect two types of abnormal behavior using optical flow: (1) a person running in a shopping mall, and (2) a person waving a hand in a crowd.

---

[1] Portions reprinted, with permission, from Ning Ding, Yongquan Chen, Zhi Zhong, Yangsheng Xu, Energy-based surveillance systems for ATM machines, *2010 8th World Congress on Intelligent Control and Automation (WCICA)*. ©[2010] IEEE.

[2] Portions reprinted, with permission, from Jinnian Guo, Xinyu Wu, Zhi Zhong, Shiqi Yu, Yangsheng Xu and Jianwei Zhang, An intelligent surveillance system based on RANSAC algorithm, *International Conference on Mechatronics and Automation*. ©[2009] IEEE.

[3] Portions reprinted, with permission, from Zhi Zhong, Weizhong Ye, Ming Yang, and Yangsheng Xu, Crowd Energy and Feature Analysis, *Proceedings of the 2007 IEEE International Conference on Integration Technology*. ©[2007] IEEE.

[4] Portions reprinted, with permission, from Zhi Zhong, Ming Yang, Shengshu Wang, Weizhong Ye, and Yangsheng Xu, Energy Methods for Crowd Surveillance, *Proceedings of the 2007 International Conference on Information Acquisition*. ©[2007] IEEE.

H. Qian et al., *Intelligent Surveillance Systems*, Intelligent Systems, Control and Automation: Science and Engineering 51, DOI 10.1007/978-94-007-1137-2_8,

The detection of a person running in a shopping mall is very difficult using the ordinary foreground subtraction method. It is very difficult to obtain the background image in a crowded shopping mall, and even if the background is available, segmentation is extremely difficult.

However, using the optical flow method, we can obtain the velocity of each pixel, and then the revised algorithm is used to compensate for the distortion of the velocity image. It is very important to compensate for the perspective distortion of the velocity image. In the camera image, the people near the camera will be bigger than the people further away from it, and thus the velocity of the people near the camera will be faster than their real velocity if the distortion is not revised.

The velocity image is then filtered by a 3×3 template, and a threshold is set to obtain the binary velocity image. To remove the noise, the pixels of the binary image are grouped into blobs. Blobs with small areas are removed. Figure 8.1 shows how the system detects a person running in a shopping mall. The red circles in the pictures hereinafter are used to highlight the abnormal region.

We also use the optical flow method to detect a person waving a hand in a crowd. Figure 8.2 shows the three original consecutive images and three binary images obtained by the optical flow method. Periodicity analysis [162] is applied to the images shown in Figure 8.3, and the hand of the person is detected after two periods.

In very crowded regions in which extreme overlapping occurs, obtaining the background and successful segmentation are difficult. In such cases, the optical flow approach is effective for the detection of abnormal behavior.

8.2 Energy-based Behavior Analysis of Groups in Crowds

We have elaborated the estimation of crowd density in the previous chapter. However, density estimation is a static issue, whereas the detection of crowd abnormalities is a dynamic one. The latter involves factors such as direction, velocity, and acceleration, which are too important to be ignored.

In this section, we address two energy methods, which are based on intensity variation and motion features, respectively. The results of wavelet analysis of the energy curves show that both methods can satisfactorily deal with crowd modeling and real-time surveillance. A comparison of the two methods is then made in a real environment using a metro surveillance system.

To tackle crowd density problems, researchers have tried to establish precise models to build virtual environments in which crowd situations are simulated [134][147][154]. However, pedestrian kinetics in a virtual environment is completely different from that in the real world, and the image or video quality generated in a virtual environment is always much better than that obtained from a real one. Because of these limitations and the tremendous amount of computation involved, such a method is not useful in actual video surveillance systems. The second method is tracking, in which the system tries to acquire people's trajectories by keeping track of individuals as they enter the monitored scene to detect abnormalities [133][137]. This method has been proven effective for the surveillance of abnormal individuals

(a) Frame 335.

(b) Frame 343.

(c) Frame 350.

Fig. 8.1 A person running in a shopping mall.

[135][143]. However, it is not so effective in terms of accuracy and real-time operation in crowded situations where there is great amount of overlapping among pedestrians. A third method to estimate crowd density is based on texture [146][150].

In contrast to the foregoing studies, we adopt energy methods. The basic idea behind this is simple and obvious: energy exists in motion. From this starting point and based on the assumption that every pedestrian or observed object in the surveillance scene is moving in a velocity within one quantitative grade, crowd motion means greater energy. After a careful analysis of the existing video surveillance

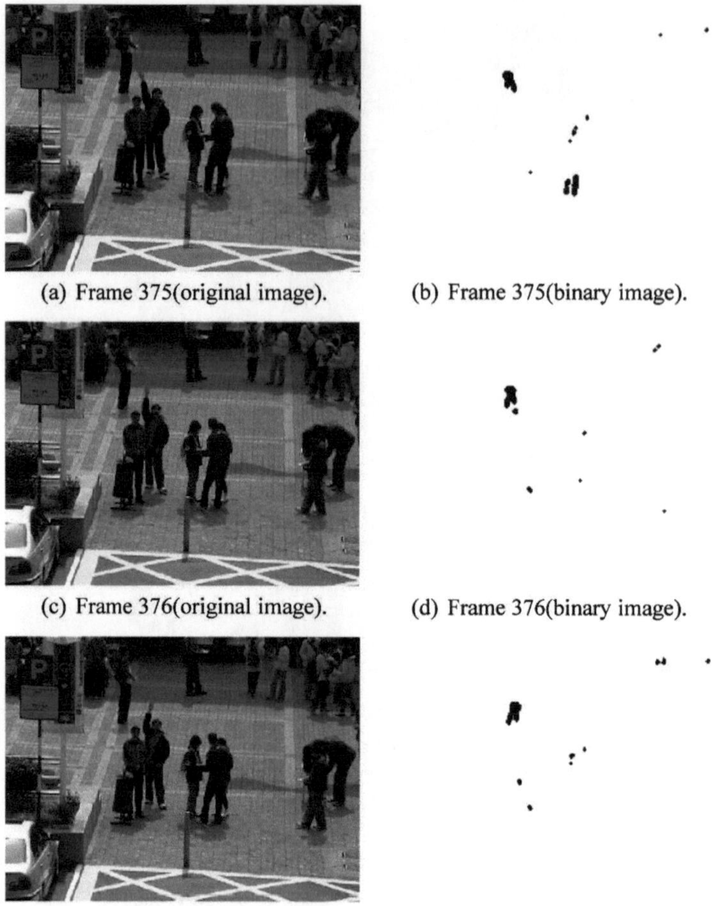

(a) Frame 375(original image). (b) Frame 375(binary image).

(c) Frame 376(original image). (d) Frame 376(binary image).

(e) Frame 377(original image). (f) Frame 377(binary image).

Fig. 8.2 Three consecutive images of people waving hands.

approaches, we define two new energy models and address two energy approaches: one is based on intensity variation, which we call "first video energy", and the other, called "second video energy", is based on motion features [156].

When we obtain the energy curves using these two methods, we can solve the previous problems. Our approaches concern the overall monitoring of the state of pedestrians in the scene and the analysis of information about their movements. Wavelet analysis is capable of simultaneously providing time and frequency information, which gives a time-frequency representation of the signal. It can be used to analyze time series that contain nonstationary power at many different frequencies [138]. When we regard crowd energy as a 1D digital signal and adopt the discrete wavelet transform (DWT) for analysis, we can get wonderful results. As we shall see in the latter part of the chapter, two kinds of abnormalities hidden in crowd energy can be detected and separated using DWT multiple-level decomposition technology [145].

Fig. 8.3 A person waving hand in a crowd.

8.2.1 First Video Energy

The energy approach that we propose for video surveillance is based on a fundamental principle: wherever there is motion, there is energy. This type of energy is often referred to as kinetic energy. Assuming that the rate of velocity of a pedestrian or an object such as a vehicle is set within a certain order of magnitude, it is clear that any motion with a larger crowd must represent more energy. Now, the remaining problems are how to define the form of the energy, how to obtain the energy, and how to carry out crowd surveillance using the parameters acquired. Viewed from the two perspectives of intensity variation and optical flow motion features, two different definitions of video energy are put forward and their applications are respectively discussed in this section.

8.2.1.1 Definition of First Video Energy

The phrase "first video energy" reflects a computer perspective, and the term "video energy" is suggested to distinguish this type of energy from that used to describe a static image. In the field of image processing, energy denotes a statistical measure of an image, which is an image texture [141]. This definition of energy better suits a partial and static description of an image, but does not suit an overall and dynamic description of video [148].

A computer takes a single-frame digital image (e.g., a video frame in a frame sequence) as 3D information, x, y, I, in which I is fixed for a single frame, while a digital video is stored in a computer as a matrix whose intensity (RGB value or I value in HSI [hue, saturation, intensity]) varies with time t. Thus, it can be seen as 4D information $(x, y, I(t), t)$, in which $I(t)$ is a function that varies with t. As for the video format where the frame rate is constant, t is equal to the n-th frame. Therefore, $I(t)$ can also be written as $I(n)$. The fundamental difference between image and video analysis lies in the time variability. In crowd surveillance analysis, or more generally, motion analysis, motion information is encapsulated in the variable $I(n)$. Broadly speaking, the kinetic formula for a particle movement system is

$$E = \sum_{i=1}^{N} m_i v_i^2 \tag{8.1}$$

The change rate of each pixel in $I(n)$ is represented by the intensity difference between two consecutive frames of that pixel. If the value of $\Delta I(n)$ is big enough, then it will be marked as being active, where the change rate of each pixel is equal to the velocity of the particle. Thus, by imitating Equation (8.1), we can get the equation for first video energy as follows:

$$VE1(n) = \sum_{i=1}^{N} w\Delta I(n)^2 \tag{8.2}$$

where N is the sum of the pixels in active blocks, and w denotes the weight of the blocks. As we do not differentiate pixels one by one, w here is the quantity of the pixels within the blocks.

Figure 8.4 shows how first video energy describes motion and crowd status in a video clip. The three lower pictures are difference frames, and the colored stick with a ball head represents the energy of that block of pixels. Different colors indicate the different motion energies of blocks, while the height of the stick represents the quantity of kinetic energy of that block.

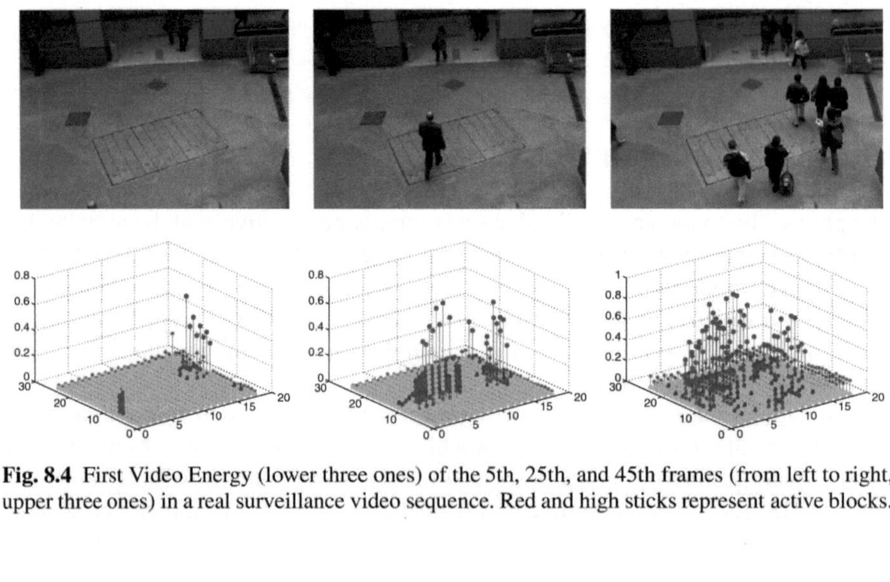

Fig. 8.4 First Video Energy (lower three ones) of the 5th, 25th, and 45th frames (from left to right, upper three ones) in a real surveillance video sequence. Red and high sticks represent active blocks.

As for the actual surveillance scene, we do not need to analyze the variation in each pixel, because doing so would be not only inefficient but also unnecessary. Quartation is exploited to segment difference frames, which creates a mosaic system. This is a recursive process, in which the final number for quartation is obtained statistically, and is discussed in the next subsection.

In the metro surveillance system in Hong Kong, the frame size is 640×480 (this is also the most popular frame mode on the market). After $K = 5$ times of quartation, the difference frame is divided into smaller blocks, each of which is $20 * 15$. Hence, the number of blocks is $N = 32 * 32$ in Equation 8.2. The adopted storage mode is HSI, so $I(t)$ is equal to I of HSI [155]. Figure 8.11 shows the curve of first video energy. Section 4 explains how it is used to monitor crowd scenes.

8.2.1.2 Quartation Algorithm

In computing first video energy, the size of the smallest possible block is the key parameter. In essence, it is an issue of adaptive image segmentation, which has been the object of much research and for which a number of algorithms have been proposed [136][142]. However, the sliding window or any other standard technique of image segmentation involves the use of empirical strategies, whereas in the recursive quartation method, the number of iterations and the proper block size are determined by a statistical approach, the F test.

The assumptions of the quartation algorithm are as follows. 1) The intensity of pixels is a stochastic variable and satisfies normal distribution. For most pixels, their mean is related to the quality of the background. 2) For every quartered block, the average pixel intensity is a stochastic variable and satisfies normal distribution. The latter assumption can easily be deduced from the former. In the former assumption, four more random variables (the means of four new blocks) appear after one former quartation.

If the quartation algorithm goes on without intensity inhomogeneity, then the existing segmentation can satisfy the need for more precise energy computation, which is the termination condition for this recursive algorithm. That is why the segmentation of each pixel is unnecessary.

The quartation algorithm is described as follows.

STEP 1: Calculate whole variance

As shown in Figure 8.5[5], the difference frame is divided into four blocks, A, B, C, and D, each of which is 320×240. For convenience, we use the letters A, B, C, and D to denote the average intensities of the blocks. If $M = (A+B+C+D)/4$, then the whole variance of the difference frame is

$$Var = \frac{1}{N-1}\sum_{j=1}^{N}(x_j - M)^2 \tag{8.3}$$

[5] Zhi Zhong, Ning Ding, Xinyu Wu, and Yangsheng Xu, Crowd Surveillance using Markov Random Fields, *Proceedings of the IEEE International Conference on Automation and Logistics*, pp. 1822-1828 © [2008] IEEE

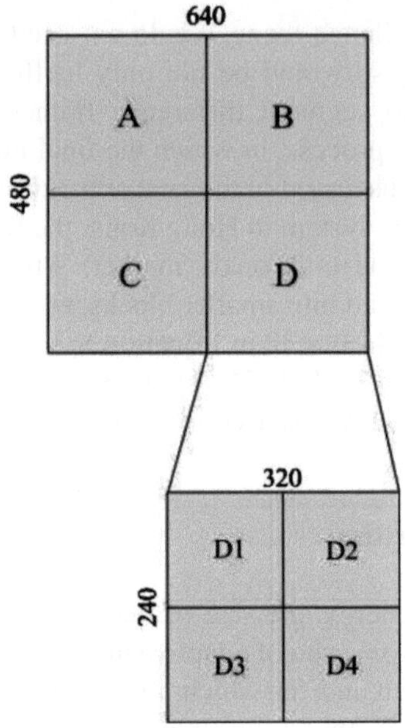

Fig. 8.5 Recursion step of Quartation Algorithm

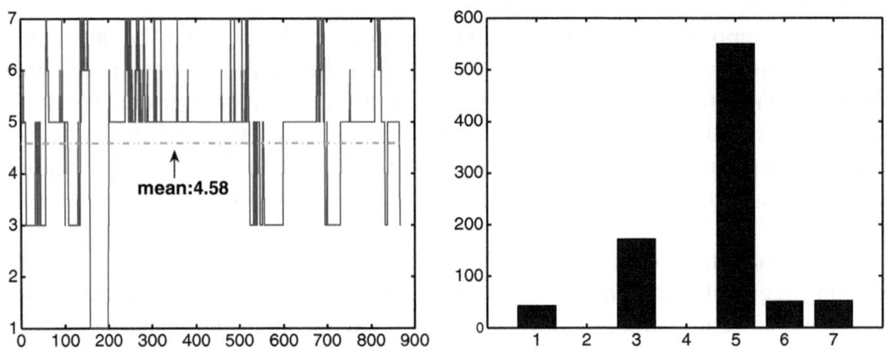

Fig. 8.6 Statistical distribution of $k(n)$. There are 868 frames. Left: $k(n)$ mostly between 3 and 7, mean is 4.58, Right: $k(n)$ of 550 frames is 5.

In (3), $N = 4$, and $x1, x2, x3$, and $x4$ are A, B, C, and D, respectively.

STEP 2: Calculate variances of blocks

Quartation is performed on A, B, C, and D, respectively. For example, in Figure 8.5, block D is further divided into four smaller blocks, which are marked D1, D2, D3, and D4. The variance of blocks A to D, $Var(A)$ to $Var(D)$, can then be computed.

STEP 3: F test

An F test is any statistical test in which the test statistic has an F distribution if the null hypothesis is true. Here, we use F to record the variance ratio of the two samples. If it is far smaller than 1 (or far greater than 1), then the two samples have very strong inhomogeneity [149]. In this chapter, A, B, C, and D and D1, D2, D3, and D4 can be treated as two samples with different variances. Then, F is either $Var/Var(D)$ or $Var(D)/Var$, depending on which value is smaller than 1. Suppose that

$$F(D) = \frac{Var(D)}{Var} \tag{8.4}$$

Then, we can get F(A), F(B), and F(C). Suppose that F(D) is the smallest term. That means that among all blocks, block D has the greatest intensity inhomogeneity in the whole difference frame.

STEP 4: Recursion

If F(D) is less than a fixed threshold, then the inhomogeneity of block D and the whole difference frame changes greatly, and quartation is repeated on block D. If F(D) is bigger than the threshold, then the intensity tends to be homogeneous, and quartation is terminated. The time of executing quartation, k, is then recorded.

STEP 5: Mosaic

The whole difference frame is mosaicked into 2^k blocks, and the mean intensity of every block is calculated.

After Step 5 of the quartation algorithm, first video energy can be calculated, and the curve can be found. Theoretically speaking, the quartation algorithm finds a local optimum. However, after analyzing k statistically, we find that it can satisfy the precision requirement. To improve the stability and the efficiency of the algorithm, we analyze the statistical distribution of $k(n)$ of a video captured by a real surveillance system. The results are shown in Figure 8.6, in which the recursion time is $K = 5$.

8.2.2 Second Video Energy

Second video energy is another way to model the motion energy of the video and to detect crowd abnormalities. This method is based on motion features [156], which can represent the motion status of the scene under surveillance.

8.2.2.1 Motion Feature

In the narrow sense, features refer to the texture of an image [151]. With the development of data mining technology, features in the broad sense are specifically defined as measurements of one or a sequence of images [152]. Here, the Harris corner is adopted as the static feature [140, 156].

The motion feature described in this paper is obtained through the image feature related to optical flow. For contrast, the latter is named the static feature hereafter. In this paper, the corner information is selected as the static feature. There are two reasons for this choice. First, we want to make sure that the proposed system works

Fig. 8.7 Frames of a railway station surveillance system. Readers can find overlapping lines increasing from left to right, so does corner information (see [139]).

in real time. Here, we do not adopt the feature in its broad sense, as in most cases, it is a recessive calculation and therefore requires much calculation time. Second, the texture in a cluttered environment, for the most part, is represented in the form of the background, which does not reflect the actual situation of people in motion. As is shown in Figure 8.7, because of the many outlines of people themselves and their overlapping contours, there exist a large number of corner features that can easily be captured and tracked.

This paper adopts the Harris corner as a feature [140]. The Harris corner is based on the local autocorrelation function, which measures the local changes in the image with patches that have shifted by a small amount in different directions. Using a truncated Taylor expansion, the local autocorrelation function $c(x,y)$ can be approximated as 8.5

$$c(x,y) = [\Delta x \Delta y] C(x,y) \begin{bmatrix} \Delta x \\ \Delta y \end{bmatrix} \tag{8.5}$$

In 8.5,

$$C(x,y) = \begin{bmatrix} \Sigma_W (I_x(x_i,y_i))^2 & \Sigma_W I_x(x_i,y_i) I_y(x_i,y_i) \\ \Sigma_W I_x(x_i,y_i) I_y(x_i,y_i) & \Sigma_W (I_y(x_i,y_i))^2 \end{bmatrix} \tag{8.6}$$

where $C(x,y)$ captures the intensity structure of the local neighborhood. Let λ_1, and λ_2 are the eigenvalues of matrix $C(x,y)$. If both eigenvalues are high, then the local autocorrelation function is sharply peaked, and shifts in any direction will result in a significant increase. This indicates a corner [111].

It is necessary to calculate $C(x,y)$ at every pixel and mark corners where the quantity Mc exceeds some threshold. M_c is defined as

$$M_c = \lambda_1 \lambda_2 - \kappa(\lambda_1 + \lambda_2)^2 \tag{8.7}$$

It can be derived that

$$M_c = \det(C(x,y)) - \kappa \cdot \mathrm{trace}^2(C(x,y)) \tag{8.8}$$

$\kappa \approx 0.04$ makes the detector a little edge-phobic, which overcomes the staircase effect, where corner detectors respond to discredited edges. In the literature, values in the range of $[0.04, 0.15]$ have been reported as feasible [111].

By arranging M_c in order of its magnitudes, we obtain a set of static features of a specific number of a frame (Harris corners). Static features by nature give the location information of key pixels. The motion features are obtained by tracking the static features of a series of images using the Lucas-Kanade optical flow approach.

Optical flow methods aim to calculate the motion fields between two image frames. In this paper, we do not need to compute the whole region of motion as this conventional method takes much computation. Instead, as shown in the table below, we use the Lucas-Kanade optical flow approach [113] to track the static features of a group of three successive frames, to determine the parameters (shown in Table 8.1) of a set of feature points whose position is specified by the static features. The feature described above is called the motion feature because it is related to the motion of the object being tracked. As can be seen from the following section, each motion feature can be regarded as a moving unit for the estimation of crowd energy.

Motion features are obtained by tracking the corners of a video sequence using the Lucas-Kanade optical flow approach [144]. Our approach tracks the corners of a group of three successive frames to determine the parameters (shown in Table 8.1) of a set of motion features.

Table 8.1 Parameters of motion feature.

Symbol	Quantity
x	horizontal position of Motion Feature
y	vertical position of Motion Feature
Δx	derivative of x
Δy	derivative of y
v	magnitude of the velocity
Angle	orientation of the velocity
Δ Angle	difference of the orientation
d^2x	derivative of dx
d^2y	derivative of dy
a	magnitude of the acceleration
Angle2	orientation of the acceleration

8.2.2.2 Definition of Second Video Energy

In the field of image processing, energy is a statistical measure used to describe the image, which is an image texture [114]. This definition of energy suits a partial, static description of an image, but does not suit an overall dynamic description of video [115].

Our definition of energy does not concern the corresponding relation of all of the points, as they cannot possibly be tracked all of the time. Even if there were the possibility of doing so, it would involve an enormous amount of calculation. Our definition is more holistic. We are concerned about whether the movement in the monitored scene is normal or abnormal. For this purpose, we define crowd energy based on the concept of kinetic energy to describe the crowdedness of an environment. The crowd energy of each frame of a picture is defined as follows:

$$VE2(n) = \sum_{i=1}^{m} w_i(n) v_i^2(n) \tag{8.9}$$

where $w_i(n)$ and $v_i(n)$ are respectively the weight and velocity of the ith point of the motion feature. Altogether, the frame has m points of the motion feature.

This way, we can obtain the value of the energy function for every video frame, which can be used to monitor the scene for abnormalities.

It is necessary to make clear the meaning of w_i. First, we take into consideration only the sum of the velocity squared, i.e., $w_i = 1$. Two problems arise. First, as can be seen from Figure 8.7, at a given time, the number of motion features of persons differs. Further, at different times, the number of motion features of the same person also varies. Hence, the physical implication represented by motion features is not the same. Therefore, each motion feature should be given a weight. Second, when overlapping takes place in a monitored scene, the number of motion features tends to be smaller than in a scene in which persons are sparsely scattered. However, in actual fact, the overlapping of persons indicates a crowded environment (in the next section, we show that this phenomenon is related to the first category of crowd abnormality that we have defined). Therefore, under such circumstances, a greater weight should be given to motion features in areas in which overlapping occurs and a smaller weight be assigned to more isolated motion features, as they are usually irrelevant to pedestrians.

We define w_i for the ith motion feature as follows. Within NN (number of neighbors) space neighbors (the space neighbors of a motion feature are defined by motion features whose Euclidean distance d is no more than $d0$ pixels, i.e., $d \leq d_0$) of a motion feature, if any one of them also has a similar neighbor (which has a greater similarity to the ith motion feature, i.e., $S_{ij} \geq S_0$) of the motion feature, then we can get

$$w_i = \sum_{j=0}^{NN} S_{ij}, (S_{ij} \geq S_0, S_0 \in (0,1]) \tag{8.10}$$

In the metro surveillance system introduced in this chapter, S_0 is 0.95, and d_0 is 20. Here, 20 is an experiential value. If we choose the ith motion feature as the center and $2d_0$ as the diameter, then the circle can cover the profile of a person in the metro surveillance system.

We adopt a measure similar to that introduced in [116] to calculate the similarity between two motion features, which is defined as

$$S_T(\overline{x}, \overline{y}) = \frac{\overline{x}^T \overline{y}}{\|\overline{x}\|^2 + \|\overline{y}\|^2 - \overline{x}^T \overline{y}} \tag{8.11}$$

The similarity measure adopted here does not require the normalization of the two comparative parameters $\overline{x}$, $\overline{y}$, although many others require normalization, such as the inner product. This measure can be used for real- and discrete-valued vectors, which makes it computationally easy and useful for extension in future work. After adding or subtracting $\overline{x}^T\overline{y}$ in the denominator of the foregoing equation, and by other algebraic manipulations, we obtain

$$S_T(\overline{x},\overline{y}) = \frac{1}{1+\frac{(\overline{x}-\overline{y})^T(\overline{x}-\overline{y})}{(\overline{x}^T\overline{y})}} \tag{8.12}$$

From 8.12, we can see that the denominator is always larger than 1, which means $S_T(\overline{x},\overline{y}) \in (0,1]$. At the same time, the similarity measure between $\overline{x}$ and $\overline{y}$ is inversely proportional to the squared Euclidean distance between $\overline{x}$ and $\overline{y}$ divided by their inner product. $\overline{x}$ is the parameter of the ith motion feature, while $\overline{x}$ is the parameter of the neighbor of the ith motion feature.

8.2.3 Third Video Energy

A video is a sequence of varying images. The difference between proximate frames reflects in part changes in the posture and movement of objects in the field of view. Motion is a process in which energy is released and exchanged. Different human behaviors demonstrate different energy features. Therefore, video energy can be used to describe the characteristics of different kinds of human behavior in the scene under surveillance. Mining more energy from the video stream is the core of this system.

Relative to the slow pace of normal behavior, an action of unusually great speed is considered abnormal. It is clear that an object moving quickly contains more kinetic energy. Thus, the speed of motion is the feature that we need to consider. In addition, violent behavior involves complex body movements. How to describe such complexity is our essential work. Thus, a weighted kinetic energy is proposed.

8.2.3.1 Definition of Third Video Energy

We have explained that motion features can be obtained by tracking corners from a video stream through an optical flow algorithm. Recently, more accurate optical flow algorithms have been proposed. For instance, Xiao et al. [129] presented a bilateral filtering-based approach, which demonstrates a high level of accuracy in building motion fields. However, the computation time is rather long. Taking into consideration real-time requirements, we use the Lucas-Kanade [144] and Horn-Schunck optical flow [131] algorithms. We adopt energy-based methods for crowd estimation with a metro video surveillance system. Here, we review the motion features based on the optical flow motion field shown in Table 8.1.

We extract the kinetic energy from the video to estimate the degree of crowd density according to the equation below:

$$E(n) = \sum_{i=1}^{W} \sum_{j=1}^{H} w_{i,j}(n) \cdot v_{i,j}^2(n) \tag{8.13}$$

The parameter $v_{i,j}(n)$ is the velocity of the $(ith,\ jth)$ pixel in the nth frame $(Width = W, Height = H)$, and the coefficient $w_{i,j}(n)$ is obtained from the blob area of the current frame to describe the number of people in the blob. This equation has been defined and discussed in detail in [127, 128].

In this paper, the same equation is used. The coefficient of kinetic energy $w_{i,j}(n)$ should be reset for the identification of violent movements, including collision, rotation, and vibration, which occur constantly in the process of human body motion. As we are concerned with how to describe the discordance in and complexity of motion, we should analyze the angle component of the optical flow motion field.

8.2.3.2 Angle Field Analysis

The angle component in a motion field (angle field) can be deduced from the following steps.

1) Calculate the horizontal $u_{i,j}(n)$ and vertical $v_{i,j}(n)$ components in the complex form of every pixel in the current frame, $u_{i,j}(n) + j \cdot v_{i,j}(n)$.

2) Get the velocity component by the complex modulus (magnitude) of the complex matrix, $V_{i,j}(n) = |u_{i,j}(n) + j \cdot v_{i,j}(n)|$. Make a $MASK(n)$ by setting a certain threshold V_{min} for the velocity component matrix.

3) Obtain the angle component by the phase angle of the complex matrix. Use the $MASK(n)$ to obtain the final angle field, $A_{i,j}(n) = \arctan \frac{u_{i,j}(n)}{v_{i,j}(n)} \cdot MASK_{i,j}(n)$. The mask is used to reduce the amount of noise in the background.

Figure 8.14 shows that the angle separately distributes the normal and the abnormal situation. The angle is concentrated in a certain direction when the object is moving normally in the field of view. When aggressive events occur, the angle field presents the average distribution. Next, we use two coefficients to describe the angle discordance.

8.2.3.3 Weighted Coefficients Design

In our algorithm, two coefficients are adopted. The first is $\Delta Angle$, shown in Table 8.1, which indicates the difference in angle between proximate frames. The second is designed to represent the inconsistency of the angle in the current frame. To achieve this, we need a benchmark angle value. There are three kinds of angles that can be considered as the benchmark. The first is the average direction, which represents the common direction of movement of each body part. The second is the direction of motion of the object's centroid, which shows the general trend in motion of the object. The one that we adopt is the third, the direction of the pixel with the greatest speed. Because it is situated in the region featuring the most conflict, we should pay attention to it.

We name the coefficient $\Delta AngleM$, which can be obtained from the following steps.

1) Find the pixel with the greatest speed, and use its angle as the benchmark angle $\angle AngleMax$.

2) Calculate the difference of all pixels in the flow field with the $\angle AngleMax$ Eq.(8.14).

$$\Delta AngleM_{i,j}(n) = A_{i,j}(n) - \angle AngleMax \tag{8.14}$$

3) Some element in the matrix of $\Delta AngleM_{i,j}(n)$ should round into the range of $(-\pi,\pi)$. If the value of $\Delta AngleM_{i,j}$ is less than $-\pi$ or bigger than π, then add 2π or -2π to it. The $\Delta AngleM_{i,j}(n)$ is the absolute difference of the benchmark angle.

4) Before application, we should normalize the coefficients in the range of $(0,1)$. First, calculate the absolute matrix of $\Delta AngleM_{i,j}(n)$, and divide the $\Delta AngleM_{i,j}(n)$ by π.

To exaggerate the weight effect for better classification performance, we use two weight approaches, including (A) Equation (8.15):

$$w_{i,j}(n) = (1 + \frac{|\Delta Angle_{i,j}(n)|}{\pi} + \frac{|\Delta AngleM_{i,j}(n)|}{\pi})^2 \tag{8.15}$$

and (B) Equation (8.16):

$$w_{i,j}(n) = (\frac{|\Delta Angle_{i,j}(n)|}{\pi} \cdot 10)^2 + (\frac{|\Delta AngleM_{i,j}(n)|}{\pi} \cdot 10)^2 \tag{8.16}$$

Considering Equation 8.13, the weighted kinetic energy wE of the nth frame is obtained from Equation 8.17 below:

$$wE(n) = \sum_{i=1}^{W} \sum_{j=1}^{H} w_{i,j}(n) \cdot v_{i,j}(n)^2 \cdot MASK_{i,j}(n) \tag{8.17}$$

8.2.4 Experiment using a Metro Surveillance System

The surveillance of passenger flows at a certain metro station exit is taken as the case for our research. Figure 8.8 is an image from a video clip captured by the video surveillance system mounted at the metro station. Shown in Figure 8.8 in the lower left corner of the image is the crowd energy curve of the frame. To the right is an image in gray scale marked with motion features. The metro surveillance system uses the Matlab routine for 1D wavelet analysis in the Haar 5 mode when initiated.

Fig. 8.8 Interface of the metro video surveillance system.

8.2.4.1 Description of Abnormality

Before describing the tasks of the real surveillance system and evaluating the two video energy methods, we have to deal with two abnormal crowd phenomena, which have been defined in [156]. The first is called static abnormality, when the value of the crowd energy of a frame at a given time greatly exceeds its supposed mean value at that time. The second is called dynamic abnormality, when there is a sudden change (mainly a great increase) in crowd energy for a persistent period of time within a video sequence.

8.2.4.2 Wavelet Analysis

Wavelet analysis can be used to efficiently detect signals on- or offline, especially transient phenomena and longtime trends in 1D signals [145]. It is the perfect tool to analyze energy curves. Wavelets can be used to detect in real time jumping, breaking, and trends in the crowd energy curves. Usually, a dynamic abnormality is more meaningful because it always represents very abnormal crowd behavior, such as fighting, chaos, or gathering.

Below are two sets of images showing the surveillance of a static abnormality (Figure 8.9) and a dynamic one (Figure 8.10) from the DWT [138]. Here, we choose second video energy, but the discussion is the same as that for first video energy.

In Figure 8.9, the upper picture is the frame at the time of detection, the middle one is the crowd energy curve of the video clip, and the lower one is the five-level approximation decomposition curve (A5) in 1D DWT. At first, a threshold (the yellow line) is set to detect the static abnormality. This threshold does not directly use the value of the real-time crowd energy, which considers only the amplitude of the energy curve, but the impact of the energy, which is an integral concept.

Hence, the threshold is set at the five-level approximation decomposition plot (for an explanation of the choice of the five-level approximation decomposition plot, please refer to [145]). At the same time, this threshold can be adjusted to satisfy the statistical requirements, which means it is not fixed but adaptive. In a certain surveillance scene, the threshold will converge at a value.

In Figure 8.10, the upper pictures are the frames at the time of detection, the middle one is the crowd energy curve of the video clip, and the lower one is the five-level detail decomposition curve (D5) in 1D DWT [145]. From Figure 8.10, we can see clearly that the crowd energy jumps from a low value to a high one (yellow ellipses). This jump is not transient; rather, it is the result obtained from convolution integral analysis based on the comparison between the crowd energy of a series of frames and that of the foregoing ones. In effect, as can be seen from the figure, although the upper right of the picture is not crowded (with only five integral passengers), there is an obvious jump in the crowd energy when the female passenger runs through the scene. At that time, the system determines that an abnormal crowd phenomenon has taken place. Similarly, when a fight bursts out in the monitored scene, as people in the mob begin to run around, there will be an obvious increase in the crowd energy. At that time, the surveillance system will send a warning signal to the security department, which will respond immediately to prevent the fight from continuing and causing further harm.

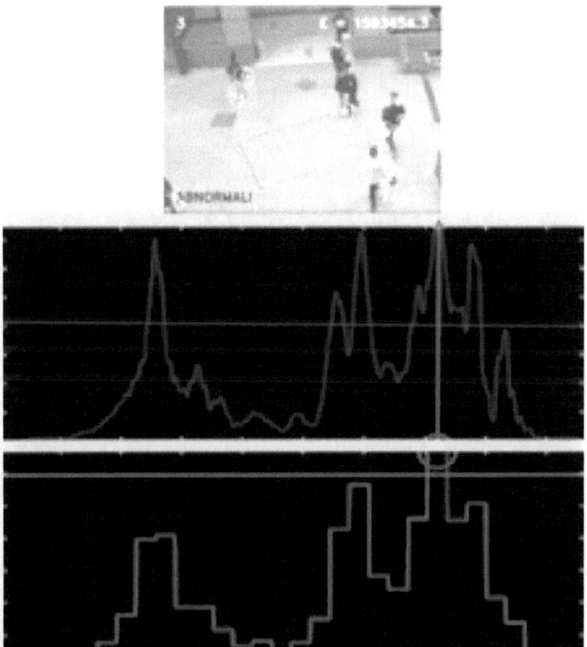

Fig. 8.9 The 5-level approximation decomposition curve (A5) in wavelet 1-D figure, from which we can see clearly that the threshold line points out the static abnormality.

8.2.4.3 Comparison of Two Kinds of Video Energy

Comparing Figure 8.11 with Figure 8.9 and Figure 8.10, respectively, we can draw the following conclusions.

First, comparing the curves of the two energies, we find that both curves have two big wave crests and one small wave crest (in the middle). The same phenomena also appear in curve D5, which exhibits a high-order property, and small wave curve A5, which exhibits a low-order property, suggesting that both video energy models are highly consistent.

Second, for the first wave crest, first video energy is greater than second video energy, whereas for the other two, the opposite is true. When carefully examining the surveillance video, we find that when the first wave crest was recorded, although there were more people, no individual was moving extraordinarily fast. However, some individuals were moving quickly when the middle wave crest was recorded (at that time, someone was running across the scene).

In addition, as shown in Figures 8.9 and 8.10, the variance and kurtosis of the second video energy curve are rather large, which, together with the explanation given in the conclusion above, suggests that second video energy is sensitive to speed and easy to detect, but not very noise resistant. First video energy, although

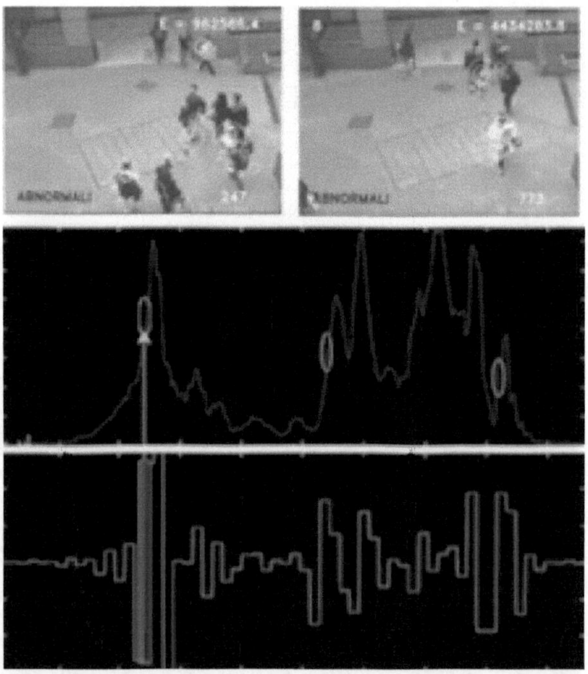

Fig. 8.10 The 5-level detail decomposition curve (D5) in wavelet 1-D figure, from which we can see the dynamic abnormality are pointed out by the yellow ellipses.

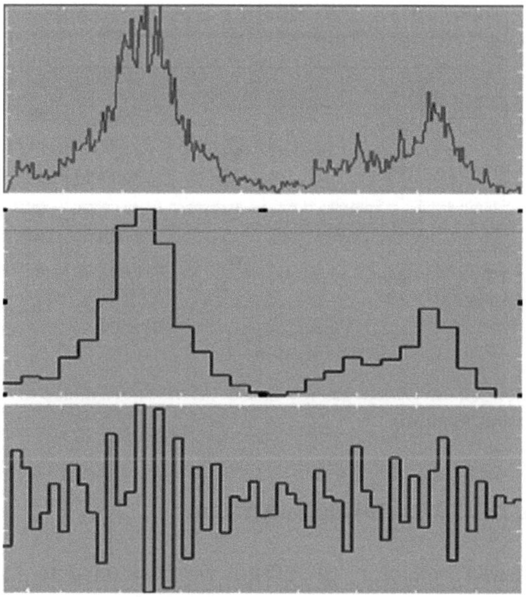

Fig. 8.11 Curve of Second Video Energy and its A5, D5 1-D wavelet analysis results.

not as speed sensitive, is rather robust, a characteristic that is confirmed by the small wave components D1, D2, D3, and D4. Theoretically, as second video energy is based on the optical flow approach, these characteristics are inherent.

8.2.5 Experiment Using an ATM Surveillance System

The framework of an ATM video surveillance system is described by the Matlab Simulink development platform. As shown in Figure 8.12, the system includes signal input, result output, and analysis subsystems. The output is used to show results and alarms. The input includes the following functions:

1) Gaussian background modeling;
2) Adaptive background updating; and
3) Video resolution resizing for expediting the processing speed.

The analysis subsystems perform the following functions:

1) Sensitive regional monitoring;
2) Monitoring for violent behavior; and
3) Logical decision making.

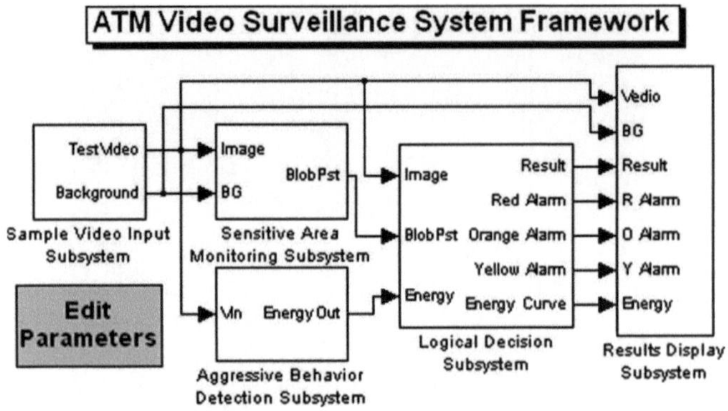

Fig. 8.12 The framework of Video Energy-based ATM Surveillance System. The system is described on Matlab and Simulink.

8.2.5.1 Sensitive Area Monitoring Subsystem

To identify abnormal behaviors at ATMs, we first have to know what normal behaviors are. Under normal circumstances, people queue up at ATMs. When someone is operating an ATM, other customers remain behind a yellow line. The strict implementation of this rule can prevent many problems from happening. Before a customer leaves the ATM booth with his card and cash, he should keep a safe distance from the others. We can set aside a region within the yellow line in front of the ATM as the sensitive area, which customers are allowed to enter one at a time. Often, more than one person at a time is standing at the ATM; for instance, couples often go to the ATM together, and children tend to stand next to their parents. To reduce the occurrence ratio of abnormal events, the drawing of a yellow line is reasonable and acceptable. With object tracking and trail analysis, the system can detect several typical kinds of abnormal behavior, including fraud, password theft, and loitering, at ATMs.

The sensitive area monitoring subsystem uses the current frame of the video and the background from the previous model as the input signal, and then exports object positions to the logical decision-making subsystem, whose functions can easily be realized with blob detection tracking. There are three steps in this subsystem:

1) Setting the region of interest (ROI);
2) Foreground extraction; and
3) Blob tracking and labeling.

8.2.5.2 Aggressive Behaviors Detection Subsystem

The real-time identification of violent acts is in high demand. However, detection methods that are based on details of the posture and body characteristics of the target do not meet the requirements [118]. In this subsystem, a four-dimensional (4D) video stream is converted into a 1D energy curve through a novel video energy mining approach, which is a dimension reduction method.

Violent acts are not confined to the region being monitored, which requires the real-time analysis of the whole scene. The input into the subsystem is the current frame, and the output is the energy value of the frame. The processing functions include:

1) Optical flow computation;

2) Mask setting;

3) Motion field and angle field extraction; and

4) Energy mining model based on the weighted kinetic energy extraction algorithm, which will be presented in detail.

8.2.5.3 Logical Decision-Making Subsystem

As Figure 8.13 shows, the logical decision-making subsystem identifies the video content based on the information from the two proceeding subsystems, and determines the corresponding level of alarm. The decision rules are as follows.

1) Violent act detection possesses the highest weight according to the feature of the energy curve.

2) The ROI has no more than one blob to prevent interference from other people waiting behind the yellow line when the customer is operating the ATM.

3) The size of the blob accords with normal human activities. This rule prevents the ATM from being operated by more than one person at a time.

4) The time that the blob stays in the area should not go beyond the threshold to prevent malicious operations.

The system defines three levels of alarm to describe the degree of abnormality: a red alarm indicates a violent event; a yellow one means a minor violation, such as crossing the yellow line or interrupting the customer operating the ATM; and an orange alarm indicates a serious violation, which lies between the normal and violent bounds. The following situations fall into the orange (warning) category:

1) Loitering in the sensitive area;

2) More than one person at the ATM or exceeding the time threshold; and

3) Dramatic fluctuation and fast increase in the energy curve.

8.2.5.4 Experiments

We randomly survey 30 ATM outlets in NanShan district in Shenzhen, a big city in south China, for installation information and experimental videos of ATM surveillance systems. The position and number of cameras depend on the shape and space of the ATM booth to meet the requirements of face identification and event recording. A realistic video dataset of robberies and other abnormal events at ATMs is rather difficult to collect. When we view and analyze clips of actual cases downloaded from the network, we find that those cases occur in different scenes, and that the position of the camera and quality of the video varies considerably with the position of camera and location. For the comparability of the experimental results, we imitate those real events and act them out in a simulated location to test our algorithm.

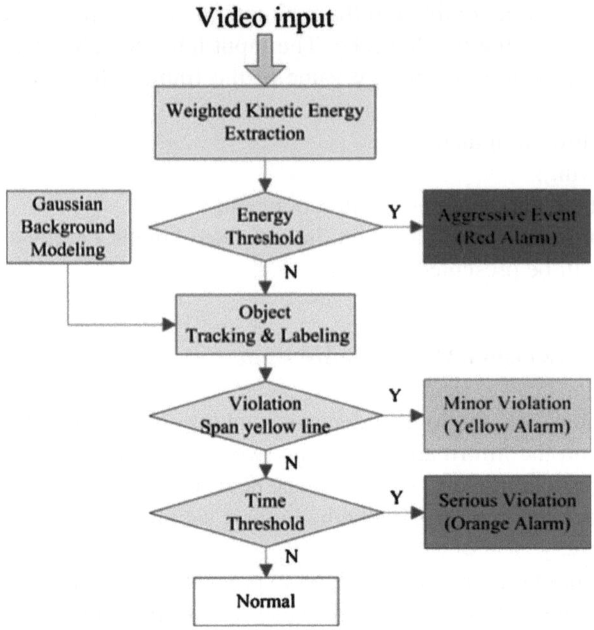

Fig. 8.13 Logical Flowchart

A traditional closed-circuit television (CCTV) video surveillance system is installed in the laboratory to simulate the ATM scene, as shown in Figure 8.16. The door of the lab is considered the location of the ATM, and a yellow line is drawn $1m$ away from the door. The camera is placed on top of the wall close to the ATM to make sure that the customer operating the machine does not block the yellow line from the camera's view. The experimental materials, which consist of 21 video clips running from 20 to 30 seconds each, contain normal situations in the first part and several typical abnormal events such as fraud, fighting, robbery, and so forth in the second part.

The details of the experiment and parameters of the system are listed in Table 8.2.

Figure 8.14 illustrates the angle distribution of the normal and abnormal situation separately. Figure 8.15 shows the performance of the weighted kinetic energy in describing aggressive behavior in a clip recording four people fighting. As the figure shows, when an abnormality occurs (around the $269th$ frame, indicated by the red line), the original kinetic energy curve [125](oKEn), the energy without weight, remains stationary, as in the previous normal situation, whereas the weighted kinetic energy curve fluctuates drastically and rises to a high value within a few frames. Two high-energy events occur around the $80th$ and $180th$ frames. The energy value of the original kinetic energy curve is close to 2500, and the value of the weighted one is 2800, giving a proportion of $(1 : 1.2)$. When a real abnormal event occurs, the proportion increases to $(1 : 3\ 6)$, which means that the weighted method can restrain the weight when a high speed but not complex motion happens, but increases the weight when a notable event occurs.

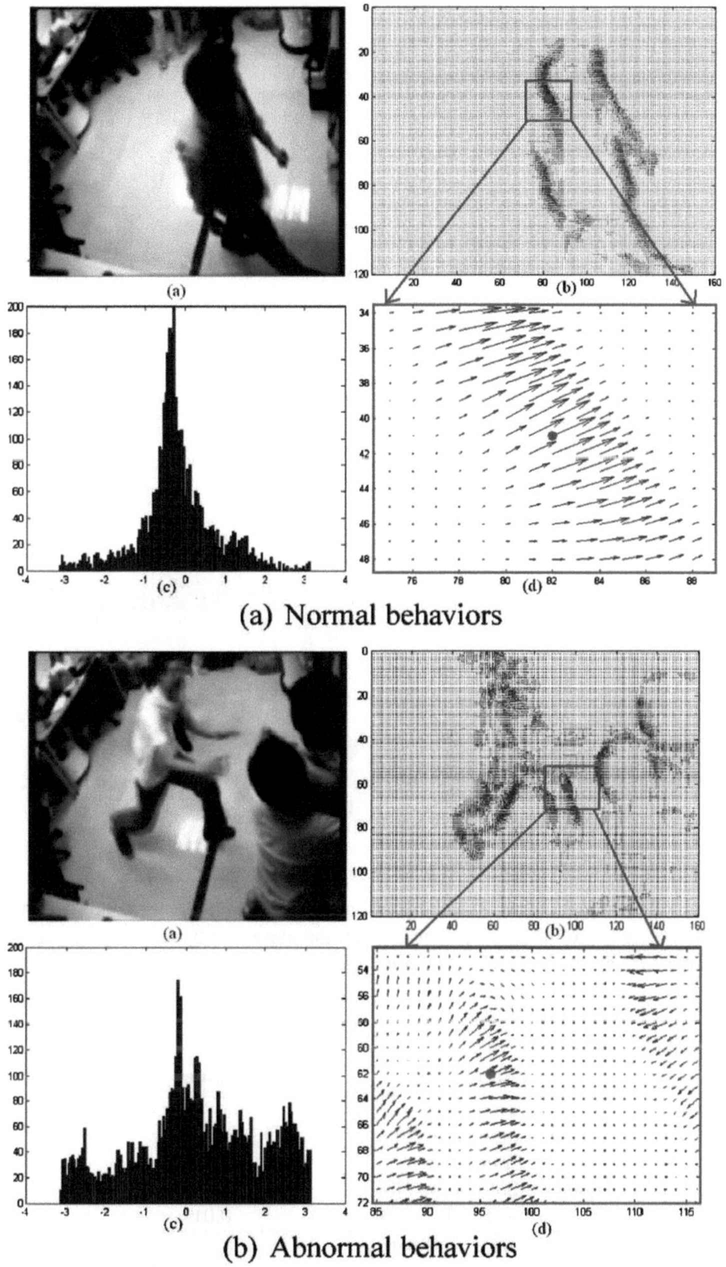

Fig. 8.14 Motion Field and Angle Distribution. The left four image is describe Normal Behaviors and the right four is Abnormal Behaviors. The sub-figure(a) is current frame. The (b) is the optical flow field. The red point in (b) indicate the pixel with highest velocity. The position mostly located in head and limbs, and change frequently along with the object posture. The (d) is the zoomed image from the red box in (b) to show detail around the red point. The (c) is the histogram of angle field.

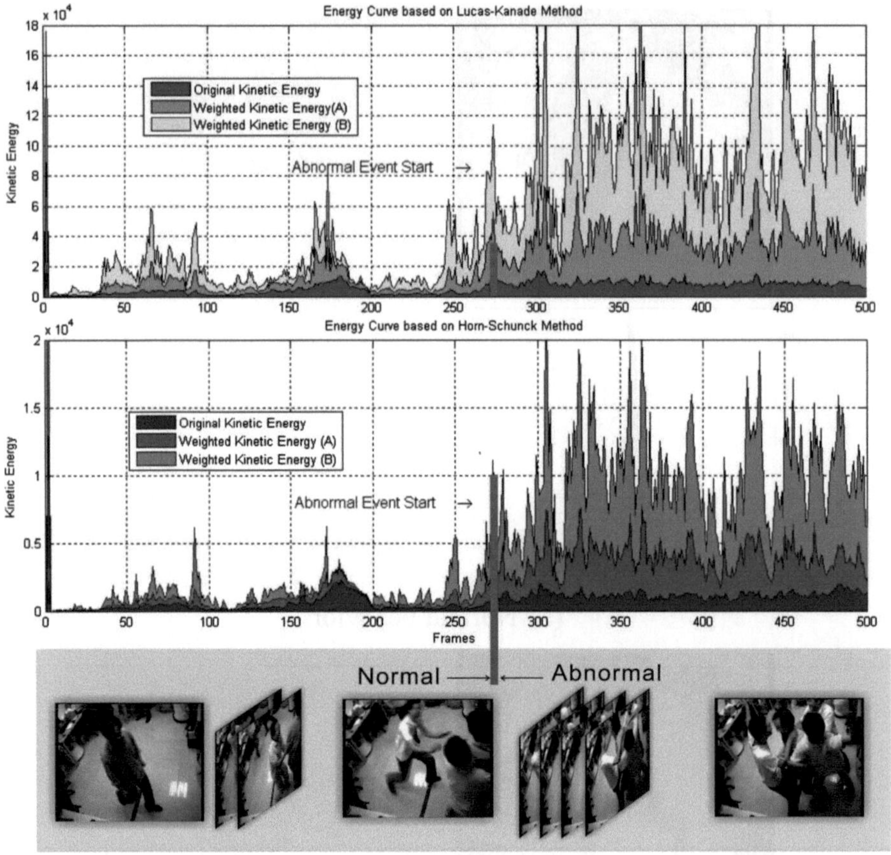

Fig. 8.15 Comparison of Energy Curves. These curves are obtained from different optical flow algorithm and weight approaches. The comparison result will help to chose the best energy extraction method. The red line indicates the time when the abnormality occurs. The sub-figure under the chart shows the snapshot of a clip contained violence.

Table 8.2 System Configuration

Parameters	Value
Video resolution	$576 \times 720pix$
Downsampled resolution	$120 \times 160pix$
Frame rate	$25fps$
Time threshold of the blob stay	$10s$
Velocity threshold for Mask	0.01
Software platform	
OS	Windows XP
Simulation environment	Matlab 2007a, Simulink
Hardware platform	
CPU	Inter D-Core 2140
Memory	DDR II 667 2G

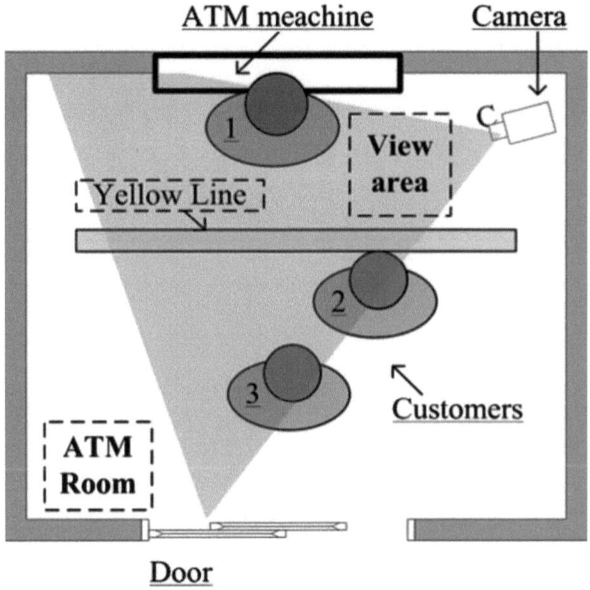

Fig. 8.16 ATM scene platform.

The difference between the weighted and the original curve is greater in the abnormal part because of the difference in the degree of the direction of motion. The subfigures (1) and (2) are the energy curves obtained using the Lucas-Kanade and Horn-Schunck methods, respectively. We choose the Horn-Schunck algorithm because those curves between the $150th\ and\ 200th$ frames prove to be rather robust in the normal part. Another comparison is made between two weighted approaches: (A) Equation (8.15) and (B) Equation (8.16). Notice that, compared to (A), the curve based on weighted approach (B) performs more robustly in the normal part and is more sensitive in the real, or abnormal, part, which is what we want. Finally, the energy extraction approach that we choose is based on the weighted method (B) and Horn-Schunck optical flow algorithm.

In addition, stationary wavelet transform (SWT) [132] is employed to analyze the energy curves, and the *sym*4 of the *Symlets* wavelet family is used to perform three-level signal decomposition. In Figure 8.17, the 3*rd*-level approximation coefficient of the energy curve shows that more energy is generated when aggressive behavior occurs. As for the 1*st*-level detail coefficient (the lowest subfigure), when a 1D variance adaptive threshold is adopted, the boundary (located in the 271*st* frame) between the two parts with different variance is quite close to the frame of the start of the abnormal event. From these results, it is clear that machine learning approaches are not required to distinguish the abnormal from the normal situation, as an energy threshold value is sufficient. In this experiment, the threshold values of the orange and red alarms are defined as 0.5×10^4 and 0.7×10^4, respectively.

Figure 8.18 shows that corresponding alarms respond to the content and the statistics of the experimental results reported in Table 8.3. The system performs with a low false negative (FN) and a high false positive (FP) rate, which means

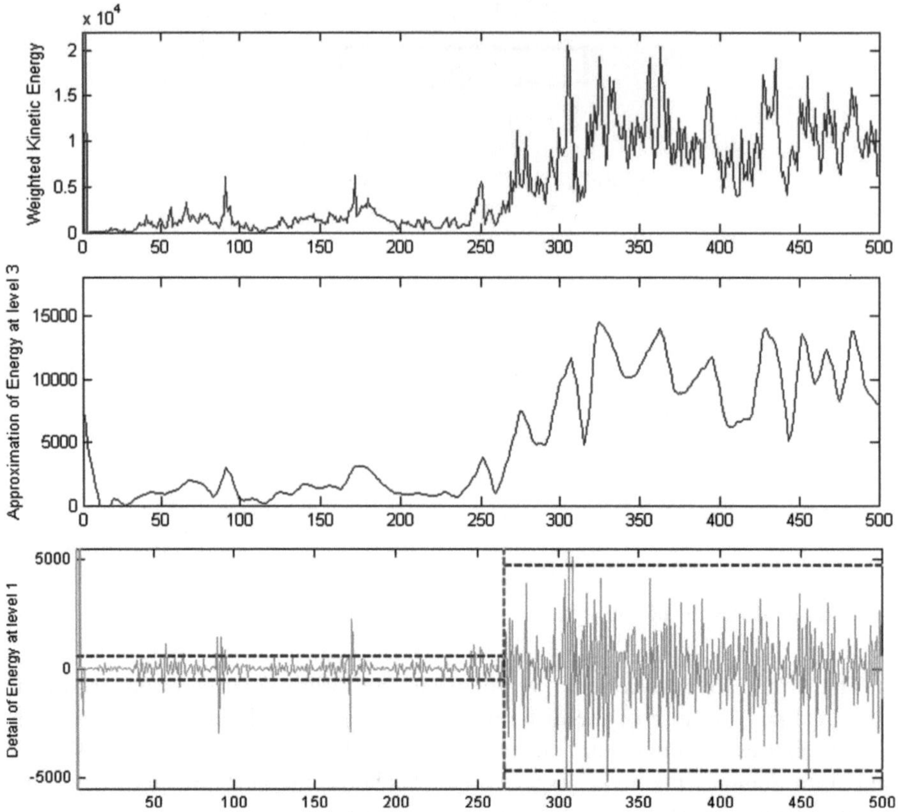

Fig. 8.17 Wavelet Analysis. The upper sub-figure is the energy curve computer by Horn-Schunck algorithm and weight approach(B). The 3-level Approximation in the *2nd* figure indicate that there is more energy when violence occur, and the 1-level detail indicate that the energy vary acutely in the abnormal part.

that it is quite sensitive to abnormal events but sometimes overreacts to normal situations. Because of the imperfections of the calibration algorithm, false positives occur mainly in the normal part of the video, when a customer walks close to the ATM. The false negative rate of nonviolent cases is lower than that of violent ones because nonviolent crime detection relies mainly on object tracking, which is more robust than the energy approach for aggressive behavior detection. The system output frame rate is around $9-11fps$, which satisfies the real-time requirement.

Table 8.3 Experimental Data

Case Type	Clips Num.	Frame Num.	FP. (%)	FN. (%)
Normal	4	2000	3.8	0
Fraud	5	3775	2.7	1.5
Fight	7	4200	2.1	1.7
Robbery	4	1700	3.6	2.3

The proposed energy-based algorithm demonstrates effective performance in solving the problem of aggressive behavior detection. The novel weighted method is not just a description of the entropy of the velocity histogram. It focuses on the pixel with the maximum velocity in the field and its relationship with other pixels, which represents an important feature of aggressive behavior. The system has been proven to be effective in the detection of nonviolent and violent behavior at ATMs. To enhance the robustness of the ATM surveillance system, our future work will concentrate on the following three aspects. First, the energy mining approach should be improved to describe abnormal situations in certain scenes. Second, a robust calibration algorithm should be considered to reduce the false positive rate. Finally, appropriate feature extraction and energy curve analysis methods would

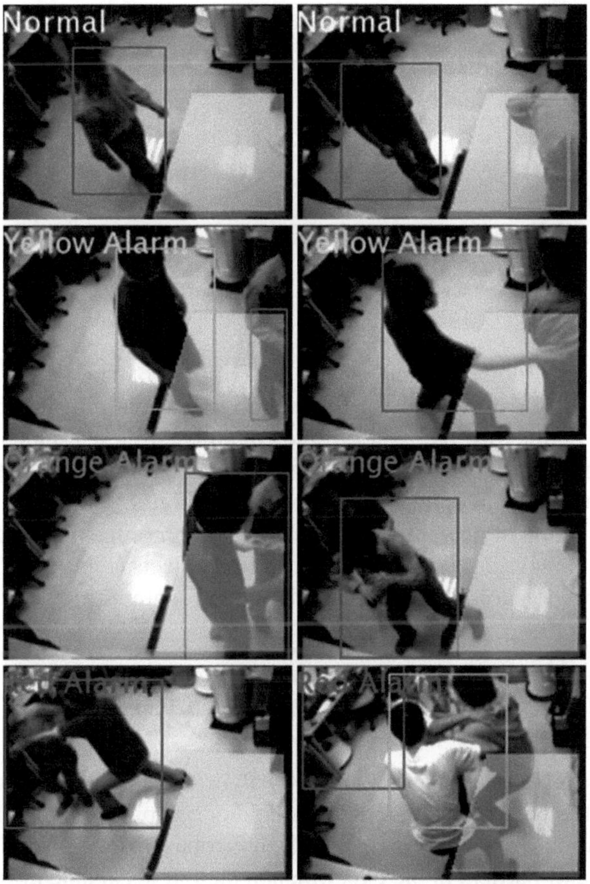

Fig. 8.18 Quad Snapshots of experimental results. The semitransparent yellow region is the sensitive area we defined at beginning. The rectangle on the people means that their motion is being tracked. The Yellow Alarm indicates that someone is striding over the yellow line when a customer is operating on the ATM. The Red Alarm warns of the occurrence of violence. The left sub-figure of Orange alarm indicate more than one customer in the area, and the right sub-figure shows the interim of violence behaviors.

provide more valuable information about abnormalities in video. The intelligence level of the system should be updated by introducing machine learning and fuzzy decision rules into complex event recognition.

8.3 RANSAC-based Behavior Analysis of Groups in Crowds

In surveillance applications in public squares, the field of view should be as large as possible. As a result, the traditional Lucas-Kanade optical flow method will introduce significant noise into the video processing.

(a) Original video frame (b) Direct optical flow of the frame

Fig. 8.19 Optical flow

To filter out the noise in images with a large field of view, we combine optical flow with foreground detection, as a foundation for various post-processing modules including object tracking, recognition, and counting. We use a statistical foreground detection algorithm that is based on a Gaussian model to reduce the level of noise pollution in the obtained optical flow, as shown in Figure 8.19.

The combination of flow information with the foreground mask allows us to consider only the flow vectors inside the foreground objects (Figure 8.20) in the analysis. The observation noise is reduced as shown in Figure 8.21.

8.3.1 Random Sample Consensus (RANSAC)

The random sample consensus (RANSAC) algorithm is a widely used robust estimator, which has become a standard in the field of computer vision. The algorithm is mainly used for the robust fitting of models. It is robust in the sense of demonstrating good tolerance of outliers in the experimental data. It is capable of interpreting and smoothing data that contain a significant percentage of gross errors. The estimate is correct only with a certain probability, because the algorithm is a randomized estimator. The algorithm has been applied to a wide range of model parameter estimation problems in computer vision, such as feature matching, registration, and the detection of geometric primitives, among others.

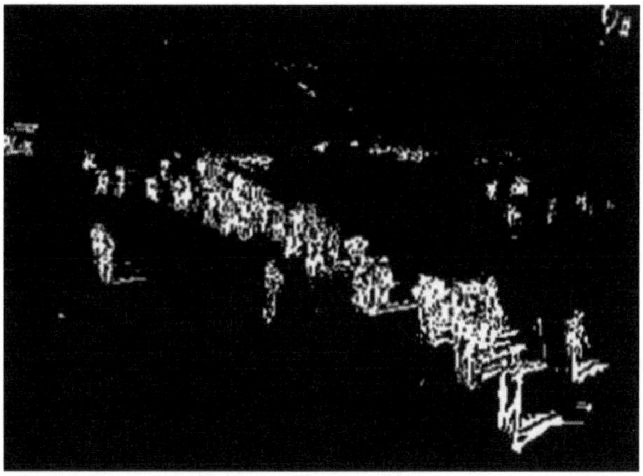

Fig. 8.20 Foreground

Fig. 8.21 Optical flow combined with foreground detection

The RANSAC algorithm easily and effectively estimates parameters of a mathematical model from a set of observed data that contains outliers. Repeatedly, it randomly selects subsets of the input data and computes the model parameters that fit the sample. It calculates the number of inliers for the model as the cost function. The process is terminated when the probability of finding a better model is lower than the probability defined by the user. The key of the problem is to deal with the outliers, which are not consistent with the supposed model. The data that are consistent with the supposed model are called inliers.

A simple example of RANSAC is shown in Figure 8.22 [157], namely, fitting a 2D line to a set of observations.

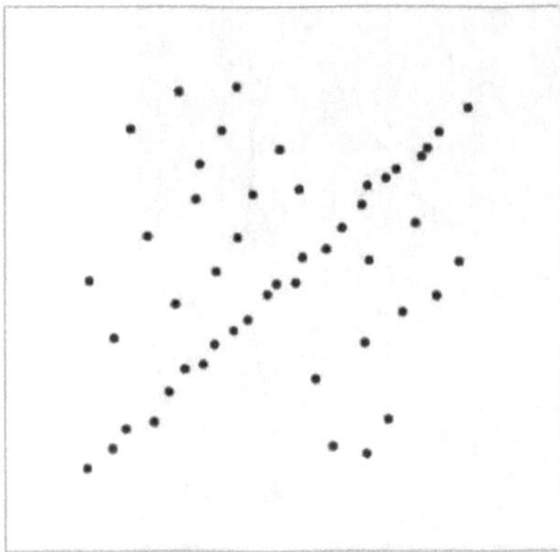

(a) A data set with many outliers to which a line has to be fitted

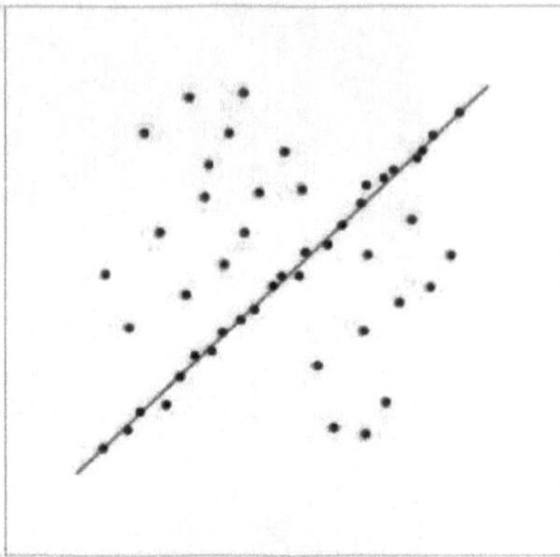

(b) Fitted line with RANSAC, outliers have no influence on the result

Fig. 8.22 An example of RANSAC

8.3.2 Estimation of Crowd Flow Direction

The abnormal behavior of crowds is always relative to the direction of crowd flow. For instance, people may walk the wrong way up a lane, or suddenly decide not to walk along the usual passage. Crowds can converge abruptly, or disperse accidentally. Here, the RANSAC algorithm is used to effectively model the direction of crowd flow.

We should define inliers and outliers in our experiments. We define one point as belonging only to inliers for the estimated model if the angle between their directions is less than δ (defined by the user); otherwise, the point belongs to outliers. The optical flow data points including outliers and inliers are shown in Figure 8.23. The yellow points are inliers while the red ones are outliers.

The details of the processing of the crowd modeling system based on RANSAC are as follows.

Input: optical flow image with the noise filtered out

Output: draw the direction of crowd flow in the image

i) Repeatedly, draw a sample from the original data. The sample here is the parameters of two optical flow points. Calculate the parameters of the direction model, which is consistent with the drawn sample.

ii) Evaluate the quality of the model. Different cost functions can be used here, according to the number of inliers.

iii) Terminating condition: the probability of finding a better model is less than h, which is defined by the user.

iv) Finally, with the inliers of the best model, estimate the parameters of the crowd flow direction model.

The structure of the algorithm based on RANSAC, M, the number of times of drawing a sample in the RANSAC algorithm, can be determined as follows.

Fig. 8.23 Optical flow data including outliers and inliers (The yellow points are inliers while the red points are outliers)

In the RANSAC algorithm, with a probability P, at least one of the M times of drawing a sample consists of inliers. Suppose that ε is the proportion of outliers in the original data, and m is the least number of points used to calculate the parameters of the model. We can then obtain the following equation:

$$1-(1-(1-\varepsilon)^m)^M = P \tag{8.18}$$

Then, we can obtain M, which is the number of times of drawing a sample, as

$$M = \frac{log(1-P)}{log(1-(1-\varepsilon)^m)} \tag{8.19}$$

The terminating condition of the algorithm: the probability of finding a better model is less than η. With a probability P, at least one of the M times of drawing a sample consists of inliers, $\eta = 1-P$.

Hence,

$$M = \frac{log(\eta)}{log(1-(1-\varepsilon)^m)} \tag{8.20}$$

We can make a table of counting M if given the dimensionality m, contamination levels ε, and confidence probability P (see Table 8.4). We define the confidence probability $P = 0.95$, $m = 2,4,6,8,10$, and $\varepsilon = 10\%,20\%,30\%,40\%,50\%$. From the table, we can conclude that M changes greatly with a change in m and ε.

Table 8.4 The Number M of Samples Required for Given ε and m $(P = 0.95)$

M	$\varepsilon = 10\%$	$\varepsilon = 20\%$	$\varepsilon = 30\%$	$\varepsilon = 40\%$	$\varepsilon = 50\%$
m = 2	2	3	5	8	11
m = 4	3	6	11	22	47
m = 6	4	19	24	63	191
m = 8	6	17	51	177	766
m = 10	7	27	105	494	3067

In our work, we aim to estimate the model parameters of the crowd flow direction. The model should be consistent, with as large a number of data points as possible. The dimensionality m used here can be defined as two. We need just two data points to calculate the model parameters. According to Table I, the number of random samples is correlated with the dimensionality m. If we define the confidence probability $P = 0.95$, then we can obtain the number of random samples as

$$M = \frac{log(1-0.95)}{log(1-(1-\varepsilon)^2)} \tag{8.21}$$

The speed of the standard RANSAC depends on two factors. The percentage of outliers determines the number of random samples needed to guarantee the 1- h confidence interval in the solution. The time needed to assess the quality of a hypothesized model is proportional to the number N of the input data points. According to 8.21 and Table I, we can conclude that the former factor will be favorable in our experiments.

To reduce the time needed for the model evaluation step, we can use the randomized RANSAC with the sequential probability ratio test (SPRT) [158]. The speed of the randomized RANSAC is increased using a two-stage procedure. First, a statistical test is performed on d randomly selected data points $(d \ll N)$. The evaluation of the remaining N-d data points is carried out only if the first d data points are inliers for the crowd flow direction model. For more details, see [159–161].

8.3.3 Definition of a Group in a Crowd (Crowd Group)

Defining what constitutes a group in a crowd (crowd group) is a pivotal task. There are at least two reasons to do so. First, it is useful for the estimation of the direction of crowd flow, for we can estimate the direction more accurately with a better crowd group definition. Second, it is the foundation for future work in estimating the density of crowd groups.

We can define a crowd group by analyzing the crowd flow direction and position, and define some people as original members of a crowd group for a specific reason. Then, we can filter the people who are inconsistent with the original crowd in two aspects: motion direction and position. The detailed operation is given as follows.

1) Based on the crowd flow direction and position, which we already know, we define the original crowd as including the people who are close to the crowd flow direction in the vertical scope.

2) Filter the people who are inconsistent with the crowd flow direction.

3) Filter the outliers in the crowd flow direction who are far away from the great mass of other people in the crowd. Let the crowd flow direction and position be expressed as the formula

$$a \times x + b \times y + c = 0 \tag{8.22}$$

For any point (x_i, y_i), the flow direction is controlled by two parameters: obliquity, θ_i, and the direction indicator, where $FLAG_i(FLAG_i) = 1$ indicates the right way, and $FLAG_i = -1$ indicates the left way. Crowd flow direction can also be expressed by the two parameters obliquity θ and direction indicator $FLAG$.

First, calculate the distance of the point to the line above. We define the point belonging to the original crowd, if the distance is less than $Limit\_a$, as

$$\frac{|a \times x + b \times y + c|}{\sqrt{a^2 + b^2}} < Limit\_a \tag{8.23}$$

Second, calculate the angle ϕ, which is between θ_i and θ, as

$$\tan\phi = \frac{|\tan\theta_i - \tan\theta|}{1 + (\tan\theta_i \times \tan\theta)} \tag{8.24}$$

We take (x_i, y_i) off the crowd if $FLAG = FLAG_i$ or ϕ is greater than $Limit\_\theta$.

For point (x_i, y_i), we calculate the distance between this and every other point (x_j, y_j). If the distance is less than $Limit\_d$, then it means that point (x_j, y_j) votes for point (x_i, y_i), which we can express as

$$STAND_{ij} = \begin{cases} 1 \ \text{if} (x_i - x_j)^2 + (y_i + y_j)^2 < Limit\_d^2 \\ 0 \ \text{otherwise} \end{cases} \tag{8.25}$$

$j = 1, 2, ..., i-1, i+1, ..., N-1, N$; N is the number of the people in the crowd. We take (x_i, y_i) off the crowd, if the proportion of the total number of affirmative votes is less than $Limit\_n$, as

$$\sum STAND_{ij} < Limit\_n \times N \tag{8.26}$$

Finally, we can define a crowd group according to the aforementioned equations. Threshold values $Limit\_a$, $Limit\_\theta$, $Limit\_d$, and $Limit\_n$ are all obtained by experiments, or can be defined by the user.

8.3.4 Experiment and Discussion

In our experiments, we aim to estimate the crowd flow direction and define a crowd group. We have run our system on a diverse set of crowd scenes. To estimate the crowd flow direction, we need to obtain the optical flow beforehand. In addition, we use the detected foreground to filter out the noise of the optical flow. Then, we use the RANSAC algorithm to estimate the parameters of the model. All of the videos used in our experiments are 320*240 pixels.

First, we run preprocessing to get the filtered optical flow. Then, we carry out the RANSAC algorithm to classify the outliers and inliers of all of the data. Next, we use the inliers to develop the model of the crowd flow direction. Figure 8.24 (a) shows the original video frame taken at the Window of the World in Shenzhen, China, and Figure 8.24 (b) shows the crowd flow direction estimated by our algorithm. Figure 8.24 (c) shows the original video frame taken at the Opening Ceremony of the Beijing Olympic Games, and Figure 8.24 (d) shows the crowd flow direction estimated by our algorithm.

(a) Original frame (b) Estimate the flow direction

(c)Original frame at Beijing Olympic Games (d) Estimate the flow direction

Fig. 8.24 Estimate the crowd flow direction with the RANSAC algorithm

We use the estimated model of the crowd flow direction and the information of the crowd position to define a crowd group. First, we filter out people who are inconsistent with the crowd flow direction. Then, we filter out the outliers in the crowd flow direction who are far away from the great mass of other people in the crowd. Figure 8.25 (a) shows one video frame taken at the Window of the World in Shenzhen, China. Figures 8.25 (b) and (c) show the semi-finished results of our algorithm. Figure 8.25 (d) shows the final results of the crowd group definition. The relative parameters are defined as follows: $Limit\_a = 30$; $Limit\_\theta = 0.35$; $Limit\_d = 50$; and $Limit\_n = 0.65$.

Combining the crowd flow direction and crowd group definition, our results are shown in Figure 8.26. Our experiments are conducted using different scenes. The overall results indicate that our algorithm can estimate the crowd flow direction and perform crowd group definition well and accurately. Figure 8.26 (a) is taken at a marketplace; Figure 8.26 (b) is taken at the Window of the World in Shenzhen; Figures 8.26 (c) and (d) are taken at Shenzhen University; and Figures 8.26 (e) and (f) are scenes of the Opening Ceremony of the Beijing Olympic Games.

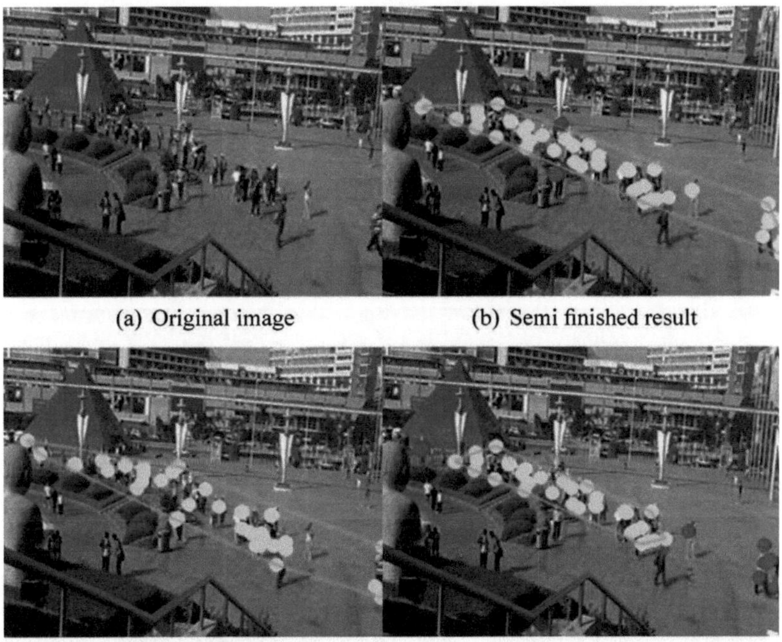

(a) Original image (b) Semi finished result

(c) Semi finished result (d) Final result

Fig. 8.25 Crowd group definition processing

(a) Some marketplace

(b) Window Of The World in Shenzhen

(c) One location at Shenzhen University

(d) Another location at Shenzhen University

(e) One scene at the Opening ceremony of the Beijing Olympic Games

(f) Another scene at the Opening ceremony of the Beijing Olympic Games

Fig. 8.26 Crowd flow direction and Crowd group definition

References

1. Yang-Jun Kim, Young-Sung Soh, "Improvement Of Background Update Method For Image Detector", 5th World Congress on Intelligent Transport Systems, 12-16. October, Seoul, Korea, 1998.
2. Andrea Prati, Ivana Mikić, Costantino Grana and etc, "Shadow Detection Algorithms For Traffic Flow Analysis: A Comparative Study", IEEE Intelligent Transportation System Conference Proceedings, Oakland (CA), USA, August 25-29, 340-345, 2001.
3. Yoshihisa Ichikura, Kajiro Watanabe, "Measurement Of Particle Flow By Optical Spatial Filtering", IMTC'94 May 10-12, Hamamatsu, 340-344, 1994.
4. David Beymer, Philip McLauchlan, Benn Coifman and etc, "A real-time computer vision system for measuring traffic parameters", Proceedings of IEEE Computer Society Conference on Computer Vision and Pattern Recognition, 495-501, 1997.
5. Stauffer C, Grimson W. E. L., "Adaptive Background Mixture Models For Real-Time Tracking[J]", IEEE Computer society Conference on Computer Vision and Pattern Recognition (Cat. No PR00149). IEEE Comput. Soc. Part Vol.2 1999.
6. C. Jiang, M.O. Ward, "Shadow Identification", Proceedings of IEEE Int'1 Conference on Computer Vision and Pattern Recognition, 1994.
7. Kenneth R. Castleman, "Digital Image Processing", Prentice-Hall International, Inc. 1996.
8. Klaus-Peter Karmann and Achim von Brandt, "Moving Object Recognition Using An Adaptive Background Memory", In V Cappellini, editor, Time- Varying Image Processing and Moving Object Recognition, 2. Elsevier, Amsterdam, The Netherlands, 1990.
9. Michael Kilger. "A Shadow Handler In A Video-Based Real-Time Traffic Monitoring System". In IEEE Workshop on Applications of Computer Vzsion, pages 1060 1066, Palm Springs, CA, 1992.
10. I. Haritaoglu, D. Harwood, and L. S Davis, "W^4: Real-time Surveillance of People and their Activities", IEEE Trans. on Pattern Analysis and Machine Intelligence, 2000.
11. Benjamin Maurin, Osama Masoud, And Nikolaos P. Papanikolopoulos,"Tracking All Traffic Computer Vision Algorithms for Monitoring Vehicles, Individuals, and Crowds", IEEE Robotics & Automation Magazine, pp 29-36, 2005.
12. T. Zhao, and R. Nevatia, "Tracking Multiple Humans in Complex Situations", IEEE Trans. on Pattern Analysis and Machine Intelligence, 2004.
13. http://www.honeywellvideo.com/products/ias/index.html.
14. http://www.eyeview.com.tw/.
15. http://www.multivision.com.hk/.
16. http://www.pixord.com/.
17. http://www.vidient.com/.
18. http://www.csst.com/about.asp .
19. M. Fathy and C. G. Darkin, "Image processing algorithms for detecting moving objects", ICARVC 94, Vol. 3, PP 1719-1723, 1994.

H. Qian et al., *Intelligent Surveillance Systems*, Intelligent Systems, Control and Automation: Science and Engineering 51, DOI 10.1007/978-94-007-1137-2,

20. N. Vaswani, A.R. Chowdhury, and R. Chellappa, “Statistical Shape Theory for Activity Modeling”, in IEEE International Conference on Acoustics, Speech, and Signal Processing, 2003.
21. X. Tao, and S. Gong, “Video Behaviour Profiling for Anomaly Detection”, IEEE Transactions on Pattern Analysis and Machine Intelligence.
22. D. Duque, H. Santos, and P. Cortez, “Prediction of Abnormal Behaviors for Intelligent Video Surveillance Systems”, in 2007 IEEE Symposium on Computational Intelligence and Data Mining, 2007.
23. H. Alghassi, S. Tafazoli, and P. Lawrence, “The Audio Surveillance Eye”, IEEE International Conference on Video and Signal Based Surveillance, 2006.
24. R. Radhakrishnan, and A. Divakaran, “Systematic Acquisition of Audio Classes for Elevator Surveillance”, in Proc. of SPIE, pp. 64-71, 2005.
25. V.-T. Vu, F. Bremond, G. Davini, et al., “Audio-Video Event Recognition System for Public Transport Security”, Institution of Engineering and Technology Conference on Crime and Security, 2006.
26. C. Clavel, T. Ehrette, and G. Richard, “Events Detection for an Audio-Based Surveillance System”, IEEE International Conference on Multimedia and Expo, 2005.
27. R.C. Luo, K.L. Su, “A Multiagent Multisensor Based Real-Time Sensory Control System for Intelligent Security Robot”, in Proceedings of the International Conference on Robotics and Automation, Taiwan, Sept, 2003.
28. A. Treptow, G. Cielniak, and T. Duckett, “Active People Recognition Using Thermal and Grey Images on a Mobile Security Robot”, in 2005 IEEE/RSJ International Conference on Intelligent Robots and Systems, 2005.
29. R.M. Haralick, “Statistical and Structural Approaches to Texture”, in Proceedings of the IEEE, vol. 67, no. 5, 1979, pp. 786–804.
30. A.N. Marana, S.A. Velastin, L.F. Costa, and R.A. Lotufo, “Automatic Estimation of Cowd Density Using Texture”, Safety Science, vol. 28, no. 3, pp. 165–175, 1998.
31. A.N. Marana, L.F. Costa, R.A. Lotufo, and S.A. Velastin, “On the Efficacy of Texture Analysis for Crowd Monitoring”, in Proceedings of the International Symposium on Computer Graphics, Image Processing, and Vision, 1998, pp. 354–361.
32. A.C. Davies, J.H. Yin, and S.A. Velanstin, “Crowd Monitoring Using Image Processing”, Electronics and Communications Engineering Journal, vol. 7, pp. 37–47, 1995.
33. C.S. Regazzoni and A. Tesei, “Distributed Data Fusion for Real-time Crowding Estimation”, Signal Processing, vol. 53, pp. 47–63, 1996.
34. C.S. Regazzoni, A. Tesei, and V. Murino, “A Real-time Vision System for Crowd Monitoring”, in Proceedings of International Conference IECON, vol. 3, 1993, pp. 1860–1864.
35. A. Tesei and C.S. Regazzoni, “Local Density Evaluation and Tracking of Multiple Objects from Complex Image Sequences”, in Proceedings of 20th International Conference IECON, vol. 2, 1994, pp. 744-748.
36. S.A. Velastin, J.H. Yin, A.C. Davies, M.A. Vicencio-Silva, R.E. Allsop, and A. Penn, “Automated Measurement of Crowd Density and Motion Using Image Processing”, in Proceedings of the 7th International Conference on Road Traffic Monitoring Contr, 1994, pp. 127-132.
37. A.N. Marana, Analise de Texturas Orientadas e Aplicacoes na Monitoracao de Multidos, Ph.D Thesis, FEEC, UNICAMP, Campinas, Brazil, 1997.
38. G.C. Lendaris and G.L. Stanley, “Diffraction Pattern Sampling for Automatic Pattern Recognition”, Proceedings of the IEEE, vol. 58, no. 2, 1970, pp. 198–216.
39. R.H. Ma, L.Y. Li, W.M. Huang, and Q. Tian, “One Pixel Count Based Crowd Density Estimation for Visual Surveillance”, in Proceedings of the 2004 IEEE Conference on Cybernetics and Intelligent Systems, 2004, pp. 170–173.
40. T. Kohonen, “The Self-organizing Map”, in Proceedings of the IEEE, vol. 78, no. 9, 1990, pp. 1464–1480.
41. S.Y. Cho, T.W.S. Chow, and C.T. Leung, “A Neural-based Crowd Estimation by Hybrid Global Learning Algorithm”, IEEE Transaction on System, Man, Cybernetics. B, vol. 29, pp. 535-541, 1999.
42. O. Chapelle, P. Haffner, and V.N. Vapnik, “Support Vector Machines for Histogram-based Image Classification”, IEEE Transaction on Neural Networks, vol. 10, pp. 1055–1064, 1999.

43. M. Oren, C. Papageorgiou, P. Sinha, E. Osuna, and T. Poggio, "Pedestrian Detection Using Wavelet Template", in Proceedings of IEEE Computer Society Conference on Computer Vision and Pattern Recognition, 1997, pp. 193-199.
44. S.F. Lin, J.Y. Chen, and H.X. Chao, "Estimation of Number of People in Crowded Scenes Using Perspective Transformation", IEEE Transaction on System, Man, Cybernetics, A, vol. 31, no. 6, pp. 645-654, 2001.
45. G. Yang and T.S. Huang, "Human Face Detection in Complex Background", Pattern Recognition, vol. 27, no. 1, pp. 53-63, 1994.
46. S. McKenna, S. Gong, and Y. Raja, "Modelling Facial Colour and Identity with Gaussian Mixtures", Pattern Recognition, vol. 31, no. 12, pp. 1883-1892, 1998.
47. A. Lanitis, C.J. Taylor, and T.F. Cootes, "An Automatic Face Identification System Using Flexible Appearance Models", Image and Vision Computing, vol. 13, no. 5, pp. 393–401, 1995.
48. Y.F. Chen, M.D. Zhang, P. Lu, and Y.S. Wang, "Differential Shape Statistical Analysis", in Proceedings of International Conference on Intelligence Computing, Hefei, China, 2005.
49. Rajagopalan, A. Kumar, K. Karlekar, J. Manivasakan, R. Patil, M. Desai, U. P. Poonacha, and S. Chaudhuri, "Finding Faces in Photographs", in Proceedings of the Sixth IEEE International Conference on Computer Vision, 1998, pp. 640–645.
50. P. Viola and M.J. Jones, "Robust Real-time Face Detection", International Journal of Computer Vision, vol. 57, no. 2, pp. 137–154, 2004.
51. H. Ren, and G. Xu, "Human Action Recognition in Smart Classroom", IEEE Proc. Int'l. Conf. on Automatic Face and Gesture Recognition, 399–404, 2002.
52. J. Ben-Arie, Z. Wang, P. Pandit, S. Rajaram, "Human Activity Recognition Using Multidimensional Indexing", IEEE Trans. on Pattern Analysis and Machine Intelligence, pp. 1091–1104, 2002.
53. K.K. Lee, and Y. Xu, "Modeling Human Actions from Learning", IEEE Int'l. Conf. on Intelligent Robot Systems, 2004.
54. I.T. Jolliffe, Principle Component Analysis, Springer-Verlag Press, Germany, 1986.
55. B. Scholkopf, and A.J. Smola, Learning with Kernels: Support Vector Machines, Regularization, Optimization, and Beyond, MIT Press, USA, 2002.
56. Y. Ou, and Y. Xu, "Learning Human Control Strategy For Dynamically Stable Robots: support vector Machine Approach", in Proc. 2003 IEEE Int'l. Conf. on Robotics and Automation, 2003.
57. D. Gutchess, M. Trajkovic, E. Cohen-Solal, D. Lyons, and A.K. Jain, "A Background Model Initialization Algorithm for Video Surveilllance", in Proc. 2001 IEEE Int'l. Conf. on Computer Vision, 2001.
58. R. Pless, J. Larson, S. Siebers, and B. Westover, "Evaluation of Local Medels of Dynamic Backgrounds", in Proc. 2003 IEEE Computer Society Conf. on Computer Vision and Pattern Recognition, 73–78, 2003.
59. http://www.activrobots.com/ROBOTS/p2dx.html.
60. R. Collins, A. Lipton, T. Kanade, H. Fujiyoshi, D. Duggins, Y. Tsin, D. Tolliver, N. Enomoto, O. Hasegawa, P. Burt, and L. Wixson, "A System for Video Surveillance and Monitoring, VSAM Final Report", Carnegie Mellon University: Technical Report CMU-RI-TR-00-12, 2000.
61. I. Pavlidis, V. Morellas, P. Tsiamyrtzis, and S. Harp, "Urban Surveillance System: From the Laboratory to the Commercial World", Proceedings of the IEEE, vol.89, no.10, pp. 1478–1497, 2001
62. A. Hampapur, L. Brown, J. Connell, S. Pankanti, A. Senior and Y. Tian, "Smart Surveillance: Applications, Technologies and Implications", in International Conference on Information, Communications and Signal Processing, vol. 2, pp. 1133-1138, 2003.
63. R. Radhakrishnan, A. Divakaran, and P. Smaragdis, "Audio Analysis for Surveillance Applications", in IEEE Workshop on Application of Signal Processing to Audio and Acoustics, pp. 158-161, 2005.

64. E. Scheirer, and M. Slaney, "Construction and Evaluation of a Robust Multifeature Speech/Music Discriminator", in Proc. of International Conference on Acoustics, Speech, and Signal Processing, pp. 1331-1334, 1997.
65. G. Tzanetakis, and P. Cook, "Musical Genre Classification of Audio Signals", IEEE Trans. on Speech and Audio Processing, vol. 10, no. 5, pp. 293–302, 2002.
66. J.N. Holmes and W.J. Holmes, Speech Synthesis and Recognition, 2nd Ed., Taylor & Francis CRC Press, 2001.
67. B.T. Logan, "Mel Frequency Cepstral Coefficients for Music Modeling", in Proceedings of the First International Symposium on Music Information Retrieval, 2001.
68. J. Foote, "An Overview of Audio Information Retrieval", Multimedia Systems, vol. 7, no. 1, pp. 2-10, 1999.
69. Y. Ou, X.Y. Wu, H.H. Qian and Y.S. Xu, "A Real Time Race Classification System", in Proceedings of the 2005 IEEE International Conference on Information Acquisition, June 27 - July 3, Hong Kong and Macau, 2005.
70. K. Fukunaga, "Introduction to Statistical Pattern Recognition", Academic Press, 1991.
71. N. Cristianini and J. Shawe-Taylor, An Introduction to Support Vector Machines and other Kernel-based Learning Methods, Cambridge University Press, Cambridge, 2000.
72. S. Bernhard, C.J.C. Burges and A. Smola, Advanced in Kernel Methods Support Vector Learning, Cambridge, MA, MIT Press, 1998.
73. http://www.grsites.com;http://www.pacdv.com.
74. T. Olson, and F. Brill, "Moving Object Detection and Event Recognition Algorithms for Smart Cameras", Proc. DARPA Image Understanding Workshop, pp. 159-175, 1997.
75. N. Massios, and F. Voorbraak, "Hierarchical Decision-theoretic Planning for Autonomous Robotic Surveillance Massios", 1999 Third European Workshop on Advanced Mobile Robots, 6-8 Sept, pp. 219-226, 1999.
76. H. Wang, D. Suter, K. Schindler and C. Shen, "TAdaptive Object Tracking Based on an Effective Appearance Filter", IEEE Trans. on Pattern Analysis and Machine Intelligence, Vol. 29, No. 9, 2007.
77. Z. Khan, T. Balch, and F. Dellaert, "MCMC-Based Particle Filtering for Tracking a Variable Number of Interacting Targets", IEEE Trans. on Pattern Analysis and Machine Intelligence, Vol. 27, No. 11, 2005.
78. J. Carpenter, P. Clifford, and P. Fernhead, "An Improved Particle Filter for Non-Linear Problems", Technical Report, Dept. Statistics, Univ. of Oxford, 1997.
79. M. Sanjeev Arulampalam, S. Maskell, N. Gordon, and T. Clapp, "A Tutorial on Particle Filters for Online Nonlinear/Non-Gaussian Bayesian Tracking", IEEE Trans. on Signal Processing, Vol. 50, No. 2, 2002.
80. V. Bruce and A. M. Burton. "Sex Discrimination: How do We Tell Difference between Nale and Female Faces?", Perception, vol. 22, pp. 131-152, 1993.
81. A.M. Burton, V. Bruce and N. Dench. "What is the Difference between Men and Women? Evidence from Facial Measurement", Perception, vol. 22, pp. 153-176, 1993.
82. B.A. Golomb, D. T Lawrence and T. J. Sejnowski, "SEXNET: A Neural Network Identifies Sex from Human Faces", Advances in Neural Information Processing Systems, pp. 572-577, 1991.
83. R. Brunelli and T. Poggio, "HyperBF Networks for Gender Classification", Proceedings of DARPA Image Understanding Workshop, pp. 311-314, 1992.
84. B. Moghaddam and M. Yang, "Gender Classification with Support Vector Machines", IEEE Intl. Conf. on Automatic Face and Gesture Recognition, pp. 306-311, 2000.
85. Z. Sun, X. Yuan and G. Bebis, "Neural-network-based Gender Classification Using Genetic Search for Eigen-feature Selection", IEEE International Joint Conference on Neural Networks, May, 2002.
86. Y.S. Ou, H.H. Qian, X.Y. Wu, and Y.S. Xu, "Real-time Surveillance Based on Human Behavior Analysis", International Journal of Information Acquisition, Vol. 2, No. 4, pp.353-365, December, 2005.
87. M. Turk and A. Pentland, "Eigenfaces for Recognition", Journal of Cognitive Neuroscience, vo. 3, no. 1, pp. 71-86, 1991.

88. P. Comon, "Independent Component Analysis, A New Concept?", Signal Processing, vol. 36, pp. 287-314, 1994.
89. A. Hyvatinen, E. Oja, L. Wang, R. Vigario, and J. Joutsensalo, "A Class of Neural Networks for Independent Component Analysis", IEEE Trans. on Neural Networks, vol. 8, no. 3, pp. 486-504, 1997.
90. M. Barlett, and T. Sejnowski. "Independent Components of Face Images: A Representation for Face Recognition", Proc. of the 4th Annual Joint Symposium on Neural Computation, Pasadena, CA, May 17, 1997.
91. G. Shakhnarovich, P. Viola and M.J. Jones, "A Unified Learning Framework for Real Time Face Detection & Classification", Proc. of the International Conference on Automatic Face & Gesture Recognition, Washington D.C., May 2002.
92. P. Viola and M.J. Jones, "Robust Real-time Object Detection", Proc. of the IEEE Workshop on Statistical and Computational Theories of Vision, 2001.
93. P.J. Phillips, H. Moon, S.A. Rizvi, and P.J. Rauss, "The FERET Evaluation Methodology for Face Recognition Algorithms", IEEE Trans. on Pattern Analysis and Machine Intelligence, vol. 22, pp. 1090-1104, 2000.
94. V. Vapnik, "The Nature of Statistical Learning Theory", Springer-Verlag, New York, 1995.
95. A. Smola, "General Cost Function for Support Vector Regression", Proceedings of the Ninth Australian Conf. on Neural Networks, pp. 79-83, 1998.
96. D.G. Lowe, "Distinctive Image Features from Scale-invariant Keypoints", International Journal of Computer Vision, vol. 60, no. 2, pp. 91-110, 2004.
97. C. Harris and M. Stephens, "Combined Corner and Edge Detector", in Proceedings of the Fourth Alvey Vision Conference, 1988, pp. 147-151.
98. Xian-Sheng Hua, Liu Wenyin and Hong-Jiang Zhang, "An Automatic Performance Evaluation Protocol for Video Text Detection Algorithms", *IEEE Transactions on Circuits and Systems for Video Technology*, Vol. 14, No. 4, April 2004, pp.498 507
99. Huiping Li, David Doermann, and Omid Kia, "Automatic Text Detection and Tracking in Digital Video", *IEEE Transactions on Image Processing*, Vol. 9, No. 1, pp. 147 156, Jan. 2000
100. Baoqing Li and Baoxin Li, "Building Pattern Classifiers Using Convolutional Neural Networks", *Proceedings of International Joint Conference on Neural Networks*, (IJCNN '99), Vol.(5), pp.3081 1085, 1999.
101. C.S. Shin, K.I. Kim, M.H. Park and H.J. KIM, "Support vector machine-based text detection in digital video", *Proceedings of the 2000 IEEE Signal Processing Society Workshop*, Vol.(2), pp. 634 641, 11-13 Dec. 2000.
102. F. LeBourgeois, "Robust Multifont OCR System from Gray Level Images", *Proceedings of the Fourth International Conference on Document Analysis and Recognition*, Vol.(1), pp. 1-5, 18-20 Aug. 1997
103. Xilin Chen, Jie Yang, Jing Zhang, and Alex Waibel, "Automatic Automatic Detection of Signs with Affine Transformation", *Proceedings of WACV2002*, Orlando, December, 2002
104. Jie Yang, Xilin Chen, Jing Zhang, Ying Zhang, and Alex Waibel, "Automatic Detection And Translation of Text from Natural Scenes", *Proceedings of ICASSP2002*, Orlando, May 2002
105. Jiang Gao, Jie Yang, "An Adaptive Algorithm for Text Detection from Natural Scenes", *IEEE Computer Society Conference on Computer Vision and Pattern Recognition Hawaii*, December 9-14, 2001
106. Jing Zhang, Xilin Chen, Jie Yang, and Alex Waibel, "A PDA-based Sign Translator", *Proceedings of ICMI2002*, Pittsburgh, October 2002
107. Ying Zhang, Bing Zhao, Jie Yang and Alex Waibel, "Automatic Sign Translation", *Proceedings of ICSLP2002*, Denver, September, 2002.
108. Y. Watanabe, Y. Okada, Yeun-Bae Kim, and T. Takeda, "Translation camera", Proceedings of the Fourteenth International Conference on Pattern Recognition, Vol. (1), pp. 613 617, 16-20 Aug. 1998
109. Trish Keaton, Sylvia M. Dominguez and Ali H. Sayed, "SNAP & TELL: A Multi-Modal Wearable Computer Interface For Browsing the Environment", Proceedings of the 6th International Symposium on Wearable Computers (ISWC'02), pp.75 82, 7-10 Oct. 2002

110. S. Theodoridis and K. Koutroumbas, *Pattern Recognition (Second Edition)*, Elsevier Science, USA, 2003
111. C. Harris, A. Blake and A. Yuille, "Geometry from Visual Motion", *Active Vision*, MIT Press, Cambridge, MA, 1992
112. K.G. Derpanis, "The Harris Corner Detector", *Technical Report*, York University
113. B. Lucas and T. Kanade, "An Iterative Image Registration Technique with an Application to Stereo Vision", *Proceedings of the International Joint Conference on Artificial Intelligence*, pp 674-679, 1981
114. K.I. Laws, "Texture energy measures", *Proc. Image Understanding Workshop*, pp 47-51, 1979
115. K.P. Ngoi and J.C. Jia, "A New Colour Image Energy for Active Contours in Natural Scenes", *Pattern Recognition Letters*, Vol. 17, pp 1271-1277, 1996
116. T. Tanimoto, "An Elementary Mathematical Theory of Classification and Prediction", *Technical Report*, IBM Corp., 1958.
117. F. Cupillard, F.Bremond, M. Thonnat. Behaviour Recognition for individuals, groups of people and crowd, *Intelligence Distributed Surveillance Systems, IEE Symposium on* , vol. 7, pp. 1-5, 2003.
118. Ankur Datta, Mubarak Shah, Niels Da, Vitoria Lobo. Person-on-Person Violence Detection in Video Data, In *Proceedings of the 16 th International Conference on Pattern Recognition.* vol. 1, pp. 433- 438, 2002.
119. Yufeng Chen, Guoyuan Liang, Ka Keung Lee, and Yangsheng Xu. Abnormal Behavior Detection by Multi-SVM-Based Bayesian Network, *Proceedings of the 2007 International Conference on Information Acquisition*, vol. 1, pp. 298-303, 2007.
120. Xinyu Wu, Yongsheng Ou, Huihuan Qian, and Yangsheng Xu. A Detection System for Human Abnormal Behavior, *IEEE International Conference on Intelligent Robot Systems*,vol. 1, no. 1, pp. 589-1593, 2005.
121. Xin Geng, Gang Li1, Yangdong Ye, Yiqing Tu, Honghua Dai. Abnormal Behavior Detection for Early Warning of Terrorist Attack, *Lecture Notes in Computer Science.* vol. 4304, pp. 1002-1009, 2006.
122. Oren Boiman, Michal Irani. Detecting Irregularities in Images and in Video, *International Journal of Computer Vision* , vol. 74, no. 1, pp. 17-31, 2007.
123. Haritaoglu, I. Harwood, D. Davis, L.S. A fast background scene modeling and maintenance for outdoor surveillance, In *Proceedings OF International Conference of Pattern Recognition*, vol. 4, pp. 179-183, 2000.
124. Jeho Nam, Masoud Alghoniemy, Ahmed H. Tewfik. Audio-Visual Content-based Violent Scene Characterization,In *Proceedings of International Conference on Image Processing (ICIP)*, vol. 1, pp. 353-357, 1998.
125. W. Zajdel, J.D. Krijnders, T. Andringa, D.M. Gavrila. CASSANDRA: audio-video sensor fusion for aggression detection., *IEEE Int. Conf. on Advanced Video and Signal based Surveillance (AVSS)*, vol. 1, pp. 200-205, 2007.
126. Thanassis Perperis, Sofia Tsekeridou. A Knowledge Engineering Approach for Complex Violence Identification in Movies, *International Federation for Information Processing(IFIP).* vol. 247, pp. 357-364, 2007.
127. Z. Zhong, W.Z. Ye, S.S. Wang, M. Yang and Y.S. Xu. Crowd Energy and Feature Analysis, *IEEE International Conference on Integreted Technology (ICIT07)*, vol.1, pp. 144-150, 2007.
128. Z. Zhong, W.Z. Ye, M. Yang, S.S. Wang and Y.S. Xu. Energy Methods for Crowd Surveillance, *IEEE International Conference on Information Acquisition (ICIA07)*, vol. 1, pp. 504-510, 2007.
129. Jiangjian Xiao, Hui Cheng, Harpreet S. Sawhney, Cen Rao, Michael A. Isnardi. Bilateral Filtering-Based Optical Flow Estimation with Occlusion Detection, ECCV (1): pp. 211-224, 2006.
130. B. Lucas and T. Kanade. An Iterative Image Registration Technique with an Application to Stereo Vision, In *Proceedings of the International Joint Conference on Artificial Intelligence*, vol. 1, pp. 674-679, 1981.

131. Barron, J.L., D.J. Fleet, S.S. Beauchemin, and T.A. Burkitt. Performance of optical flow techniques. *IEEE Computer Society Conference on Computer Vision and Pattern Recognition.* vol. 1, pp. 236-242, 1992.
132. Coifman, R.R.; Donoho, D.L., Translation invariant de-noising. *Lecture Notes in Statistics*, vol. 103, pp. 125-150, 1995.
133. E.L. Andrade, "Simulation of Crowd Problems for Computer Vision", Technical Report (EDI-INF-RR-0493), The University of Edinburgh, Available: homepages.inf.ed.ac.uk/rbf/PAPERS/vcrowds.pdf
134. S. Bandini, M.L. Federici and G. Vizzari, "A Methodology for Crowd Modelling with Situated Cellular Agents", *in WOA 2005*, Camerino, Italy, 2005, pp 91-98.
135. Y.F. Chen, Z. Zhong and Y.S. Xu, "Multi-agent Based Surveillance", *in IEEE International Conference on Intelligent Robots and Systems (IROS06)*, Beijing, China, 2006, pp 2810-2815.
136. J.Q. Chen, T.N. Pappas, A. Mojsilovic and B.E. Rogowitz, "Adaptive Perceptual Color-Texture Image Segmentation", *in IEEE Transactions on Image Processing*, VOL. 14, NO. 10, 2005, pp 1524-1536.
137. F. Cupillard, A. Avanzi, F. Bremond and M. Thonnat, "Video Understanding for Metro Surveillance", *in IEEE International Conference on Networking, Sensing and Control*, Taipei, Taiwan, 2004, pp 186-191.
138. I. Daubechies, "The wavelet transform time-frequency localization and signal analysis", *IEEE Transaction of Information Theory*, Vol.36, 1990, pp 961-1004.
139. A.C. Davies, H.Y. Jia and S.A. Velastin, "Crowd monitoring using image processing", *Electronics & Communication Engineering Journal*, Vol. 7, 1995, pp 37-47.
140. K.G. Derpanis, "The Harris Corner Detector", Technical Report, York University, Available: www.cse.yorku.ca/ kosta/CompVis_Notes/ harris_detector.pdf
141. K.I. Laws, "Texture energy measures", *in Proc. Image Understanding Workshop*, 1979, pp 47-51.
142. K.M. Lee, "Robust Adaptive Segmentation of Range Images", *IEEE Transactions on Pattern Analysis and Machine Intelligence*, VOL. 20, NO. 2, 1998, pp 200-205.
143. A.J. Lipton, "Intelligent Video Surveillance in Crowds", Technical Report, ObjectVideo Ltd., Available: www.dbvision.net/objects/pdf/ whitepapers/FlowControl.pdf
144. B. Lucas and T. Kanade, "An Iterative Image Registration Technique with an Application to Stereo Vision", *in Proceedings of the International Joint Conference on Artificial Intelligence*, 1981, pp 674-679.
145. S.G. Mallat, *A Wavelet Tour of Signal Processing*, Academic Press, London, UK, 1998.
146. A.N. Marana, S.A. Velastin, L.F. Costa and R.A.Lotufo, "Automatic Estimation of Crowd Density Using Texture", *in International Workshop on Systems and Image Processing (IWSIP97)*, Poland, 1997.
147. S.R. Musse, C. Babski, T. Capin and D. Thalmann, "Crowd Modelling in Collaborative Virtual Environments", *in ACM Symposium on Virtual Reality Software and Technology*, Taipei, Taiwan, 1998, pp 115-123.
148. K.P. Ngoi and J.C. Jia, "A New Colour Image Energy for Active Contours in Natural Scenes", *Pattern Recognition Letters*, Vol. 17, 1996, pp 1271-1277.
149. W.H. Press, W.A. Teukolsky, W.T. Vetterling and B.R. Flannery, *Numerical Recipes in C – The Art of Scientific Computing (Second Edition)*, Cambridge University Press, Cambridge, UK, 1992.
150. H. Rahmalan, M.S. Nixon and J. N. Carter, "On Crowd Density Estimation for Surveillance", *in the Conferrence of Imaging for Crime Detection and Prevention (ICDP06)*, London, UK, 2006.
151. J. Shi, C. Tomasi, "Good Feature to Track", *in IEEE Conference on Computer Vision and Pattern Recognition (CVPR94)*, Seattle, U.S., 1994.
152. S. Theodoridis and K. Koutroumbas, *Pattern Recognition (Second Edition)*, Elsevier Science, U.S., 2003.
153. T. Tanimoto, "An Elementary Mathematical theory of Classification and Prediction", Technical Report, IBM Corp., 1958.

154. B. Ulicny and D. Thalmann, "Crowd Simulation for Interactive Virtual Environments and VR Training Systems", *in Eurographic Workshop on Computer Animation and Simulation*, Manchester, UK, 2001, pp 163-170.
155. E. Welch, R. Moorhead and J. Owens, "Image processing using the HSI color space", *IEEE Proceeding of Southeastcon*, Williamsburg, USA, April 1991.
156. Z. Zhong, W.Z. Ye, S.S. Wang, M. Yang and Y.S. Xu, "Crowd Energy and Feature Analysis", *in IEEE International Conference on Integreted Technology (ICIT07)*, Shenzhen, China, 2007, pp 144-150.
157. Http://en.wikipedia.org/wiki/RANSAC
158. O. Chum and J. Matas, "Optimal Randomized RANSAC", *IEEE TRANSACTIONS ON PATTERN ANALYSIS AND MACHINE INTELLIGENCE*, VOL. 30, NO. 8, AUGUST 2008
159. O. Chum and J. Matas, "Randomized RANSAC with T(d, d) Test", *Proc. British Machine Vision Conf. (BMVC 02)*, vol. 2, BMVA, pp. 448-457, 2002
160. O. Chum, J. Matas, and J. Kittler, "Locally Optimized RANSAC", *Proc. Ann. Symp. German Assoc. for Pattern Recognition (DAGM 03)*, 2003.
161. Abraham Wald, "Sequential analysis", *Dover*, New York, 1947
162. Haritaoglu, I., Harwood, D., Davis, L.S., "W4: real-time surveillance of people and their activities", IEEE Trans. on Pattern Analysis and Machine Intelligence. Vol. 22, No. 8, August, 2000
163. Olson, T., Brill, F., "Moving object detection and event recognition algorithms for smart cameras", Proc. DARPA Image Understanding Workshop, pp. 159-175, 1997
164. Zhao, T., Nevatia R., "Tracking multiple humans in complex situations", IEEE Trans. on Pattern Analysis and Machine Intelligence, Vol. 26, No. 9, September 2004
165. Radhakrishnan, R., Divakaran, A., "Systematic acquisition of audio classes for elevator surveillance", in Proc. of SPIE, pp. 64–71, 2005
166. Luo, R.C., Su, K.L., "A Multiagent Multisensor Based Real-Time Sensory Control System for Intelligent Security Robot", in Proceedings of International conference on Robotics and Automation, Taiwan, Sept, 2003
167. Massios, N., Voorbraak, F., "Hierarchical decision-theoretic planning for autonomous robotic surveillance Massios", Advanced Mobile Robots, 1999 Third European Workshop on 6-8 Sept, 1999, Page(s):219 - 226
168. Wang, H., Suter, D., Schindler, K., Shen, C., "TAdaptive Object Tracking Based on an Effective Appearance Filter", IEEE Trans. on Pattern Analysis and Machine Intelligence, Vol. 29, No. 9, 2007
169. Khan, Z., Balch, T., Dellaert, F., "MCMC-Based Particle Filtering for Tracking a Variable Number of Interacting Targets", IEEE Trans. on Pattern Analysis and Machine Intelligence, Vol. 27, No. 11, 2005
170. Carpenter, J., Clifford, P., Fernhead, P., "An Improved Particle Filter for Non-Linear Problems", technical report, Dept. Statistics, Univ. of Oxford, 1997
171. Arulampalam, M.S., Maskell, S., Gordon N., Clapp, T., "A Tutorial on Particle Filters for Online Nonlinear/Non-Gaussian Bayesian Tracking", IEEE Trans. on Signal Processing, Vol. 50, No. 2, 2002
172. Scheirer, E., Slaney, M., "Construction and evaluation of a robust multifeature speech/music discriminator", in Proc. of International Conference on Acoustics, Speech, and Signal Processing, pp. 1331–1334, 1997
173. Tzanetakis, G., Cook, P., "Musical Genre Classification of Audio Signals", IEEE Trans. on Speech and Audio Processing, vol.10, no.5, pp. 293–302, 2002
174. Holmes, J.N., Holmes, W.J., Speech synthesis and recognition, 2nd Ed., Taylor & Francis CRC press, 2001
175. Logan B.T., "Mel frequency cepstral coefficients for music modeling", in Proceedings of the First International Symposium on Music Information Retrieval, 2001
176. Foote J., "An overview of audio information retrieval", Multimedia Systems, vol.7, no.1, pp. 2-10, 1999

177. Radhakrishnan, R., Divakaran, A., Smaragdis, P., "Audio analysis for surveillance applications", in IEEE Workshop on Application of Signal Processing to Audio and Acoustics, pp. 158–161, 2005
178. Cristianini, N., Shawe-Taylor, J., "A introduction to support vector machines and other kernel-based learning methods," *Cambridge University Press,* Cambridge, 2000
179. Bernhard, S., Burges, C.J.C., Smola, A., "Advanced in kernel methods support vector learning," *Cambridge, MA, MIT Press,* 1998
180. Ou, Y., Wu, X.Y., Qian, H.H, Xu, Y.S., "A real time race classification system", in Information Acquisition, 2005 IEEE International Conference on June 27 - July 3, 2005
181. Perez, P., Hue, C., Vermaak, J., and M. Gangnet, "Color-Based Probabilistic Tracking", European Conference on Computer Vision, pp. 661-675, 2002
182. Y.S. Ou, H.H. Qian, X.Y. Wu, and Y.S. Xu, "Real-time Surveillance Based on Human Behavior Analysis", International Journal of Information Acquisition, Vol.2, No.4, pp.353-365, December, 2005.
183. X.Y. Wu, J.Z. Qin, J. Cheng, and Y.S. Xu, "Detecting Audio Abnormal Information", The 13th International Conference on Advanced Robotics, pp.550-554, Jeju, Korea, August 21-24, 2007.
184. http://www.grsites.com
185. T. Ojala, M. Pietik*ä*inen, and T. M*ä*enp*ää*, "Multiresolution gray-scale and rotation invariant texture classification with local binary patterns", *IEEE Trasactions on Pattern Analysis and Machine Intelligence*, vol. 24, no. 7, pp. 971C987, 2002
186. G. Welch and G. Bishop, "An Introduction to the Kalman Filter", *ACM SIGGRAPH 2001 Course Notes*, 2001
187. N. Funk, "A study of the kalman filter applied to visual tracking", *University of Alberta*, 2003
188. T. Ahonen, A. Hadid, and M. Pietik*ä*inen, "Face recognition with local binary patterns", *European Conference on Computer Vision (ECCV04)*, vol. 3021, pp. 469C481, 2004
189. PETS01 datasets, http://visualsurveillance.org, The University of Reading, UK.
190. G. R. Bradski and S. Clara, "Computer vision face tracking for use in a perceptual user interface", *Intel Technology Journal*, No. Q2, pp. 15, 1998.
191. M. K. Hu, "Visual patern recognition by moment invariants", *IRE Transaction on Information Theory*, Vol. IT-8, pp. 179-187, 1962.
192. J. W. Davis and A. F. Bobick, "The representation and recognition of action using temporal templates", *MIT Media Lab Technical Report 402*, 1997.
193. G. R. Bradski and J. W. Davis, "Motion segmentation and pose recongnistion with motion history gradients", *Mach. Vision Appl.*, Vol. 13, No. 3, pp. 174-184, 2002.
194. W. Hu, T. Tan, L. Wang, and S. Maybank, "A survey on visual surveillance of object motion and behaviors", *Image and Vision Computing*, Vol. 20, pp. 905-916, 2002.
195. J. Rittscher, A. Blakeb and S. J. Roberts, "Towards the automatic analysis of complex human body motions", *Systems, Man and Cybernetics, Part C, IEEE Transactions on*, Vol. 34, No. 3, pp. 334-352, Aug., 2004.
196. C. Loscos, D. Marchal, and A. Meyer, "Intuitive crowd behaviour in dense urban environments using local laws", *Theory and Practice of Computer Graphics, Proceedings*, pp. 122-129, Jun., 2003.
197. D. B. Yang, H. H. Gonzalez-Banos and L. J. Guibas, "Counting people in crowds with a real-time network of simple image sensors", *IEEE International Conference on Computer Vision*, Vol. 1, pp. 122-129, 2003.
198. O. Frank, J. Nieto, J. Guivant, and S. Scheding, "Target tracking using sequential Monte Carlo methods and statistical data association", *IEEE/RSJ International Conference on Intelligent Robots and Systems*, Vol. 3, pp. 2718-2723, Oct., 2003.
199. P. Guha, A. Mukerjee, and K. S. Venkatesh, "Efficient occlusion handling for multiple agent tracking by reasoning with surveillance event primitives", *IEEE International Workshop on Visual Surveillance and Performance Evaluation of Tracking and Surveillance*, pp. 49-56, Oct., 2005.
200. Y. Ivanov, C. Stauffer, A. Bobick, and W. E. L. Grimson, "Video surveillance of interactions", *IEEE Workshop on Visual Surveillance*, pp. 82-89, Jun., 1999.

201. J. Orwell, S. Massey, P. Remagnino, D. Greenhill, and G. A. Jones, "A multi-agent framework for visual surveillance", *International Conference on Image Analysis and Processing*, pp. 1104-1107, Sep., 1999.
202. A. Bakhtari and B. Benhabib, "Agent-based active-vision system reconfiguration for autonomous surveillance of dynamic, multi-object environments", *International Conference on Intelligent Robots and Systems*, pp. 574-579, Aug., 2005.
203. K. Ashida, Seung-Joo Lee, J. M. Allbeck, H. Sun, N. I. Badler, and D. Metaxas, "Pedestrains: creating agent behaviors through statistical analysis of observation data", *Conference on Computer Animation*, pp. 84-92, Nov., 2001.
204. N. M. Oliver, B. Rosario, and A. P. Pentland, "A Bayesian computer vision system for modeling human interactions", *Pattern Analysis and Machine Intelligence, IEEE Transactions on*, Vol. 22, pp. 831-843, Aug. 2000.
205. P. Remagnino, T. Tan, and K. Baker, "Agent orientated annotation in model based visual surveillance", *International Conference on Computer Vision*, pp. 857-862, Jan., 1998.
206. R. Hosie, S. Venkatesh, and G. West,"Classifying and detecting group behaviour from visual surveillance data", *International Conference on Pattern Recognition*, Vol. 1, pp. 602-604, Aug., 1998.
207. C. Niederberger, and M. H. Gross, "Towards a game agent", *Institute of Scientific Computing, Zurich,Swiss*, No. 377, 2002.
208. Y. Chen, M. Zhang, P. Lu and Y., "Differential shape statistical analysis", *International Conference on Intelligence Computing*, 2005.
209. C. J. C. Burges, "A tutorial on support vector machines for pattern recognition", *Data Mining and Knowledge Discovery*, Vol. 2, No. 2, pp. 121-167, 1998.
210. L. Khoudour, T. Leclercq, J. L. Bruyelle, and A. Flancqurt, "Local camera network for surveillance in Newcastle airport", *IEE Symposium on Intelligence Distributed Surveillance Systems*, Feb., 2003.
211. Jia Hong Yin, Sergio A. Velastin, Anthony C. Davies, "Image Processing Techniques for Crowd Density Estimation Using a Reference Image", *ACCV* pp. 489-498, 1995.
212. Siu-Yeung Cho and Tommy W. S. Chow, "A neural-based crowd estimation by hybrid global learning algorithm", IEEE Transactions on Systems, Man, and Cybernetics, Part B 29(4): 535-541 (1999)
213. Sheng-Fuu Lin, Jaw-Yeh Chen, Hung-Xin Chao, "Estimation of number of people in crowded scenes using perspective transformation", *IEEE Transactions on Systems, Man, and Cybernetics, Part A*, Vol. 31, No. 6, pp. 645-654, 2001.
214. Nikos Paragios and Visvanathan Ramesh, "A MRF-based approach for real-time subway monitoring", *CVPR (1)* pp. 1034-1040, 2001.
215. Ernesto L. Andrade, Scott Blunsden and Robert B. Fisher, "Modelling crowd scenes for event detection", *ICPR (1)*, pp. 175-178, 2006.
216. Zivkovic, "Improved adaptive Gaussian mixture model for background subtraction", *ICPR2004*, Cambridge, United Kingdom, Vol. 2, pp. 28-31, 2004.
217. Huang Bufu, "Gait modeling for intelligent shoes", *IEEE International Conference on Robotics and Automation*, Roma, Italy, pp. 10-14, April, 2007.

Index

MIX
Papier aus verantwortungsvollen Quellen
Paper from responsible sources
FSC® C105338

If you have any concerns about our products,
you can contact us on
ProductSafety@springernature.com

In case Publisher is established outside the EU,
the EU authorized representative is:
Springer Nature Customer Service Center GmbH
Europaplatz 3, 69115 Heidelberg, Germany

Printed by Libri Plureos GmbH
in Hamburg, Germany